Industrial Engineering

Edited by Ainul Akmar Mokhtar
and Masdi Muhammad

Published in London, United Kingdom

IntechOpen

Supporting open minds since 2005

Industrial Engineering
http://dx.doi.org/10.5772/intechopen.75414
Edited by Ainul Akmar Mokhtar and Masdi Muhammad

Contributors
Lokpriya Gaikwad, Vivek Sunnapwar, Zahedi Zahedi, Kuldeep Chaudhary, P.C. Jha, Bolatbek Utegulov, Ikram Khatrouch, Lyes Kermad, Abderahman El Mhamedi, Younes Boujelbene

First published in London, United Kingdom, 2019 by IntechOpen
IntechOpen is the global imprint of INTECHOPEN LIMITED, registered in England and Wales, registration number: 11086078, The Shard, 25th floor, 32 London Bridge Street
London, SE19SG – United Kingdom
Printed in Croatia

British Library Cataloguing-in-Publication Data
A catalogue record for this book is available from the British Library

Additional hard copies can be obtained from orders@intechopen.com

Industrial Engineering
Edited by Ainul Akmar Mokhtar and Masdi Muhammad
p. cm.
Print ISBN 978-1-83880-345-2
Online ISBN 978-1-83880-346-9
eBook (PDF) ISBN 978-1-83880-633-0

For EU product safety concerns:
IN TECH d.o.o., Prolaz Marije Krucifikse Kozulić 3, 51000 Rijeka, Croatia, info@intechopen.com or visit our website at intechopen.com.

Meet the editors

Dr. Ainul Akmar binti Mokhtar is an associate professor in the Department of Mechanical Engineering at Universiti Teknologi Petronas (UTP), Malaysia. She is also a programme manager for the MSc programme in Asset Management and Maintenance at the university. Prior to joining UTP, she worked with Seagate Technology (M) Sdn Bhd as a process engineer. Dr. Mokhtar earned a BSc in Industrial Engineering from Purdue University, Indiana, USA, and an MSc in Manufacturing Systems from Nottingham University, England, UK. She earned a PhD in Mechanical Engineering at UTP for her research on "Failure Probability Model for Piping Systems subject to Corrosion under Insulation." She is actively involved in research and industrial projects related to reliability and maintenance, asset management, and operations research, and she supervises graduate students in these areas. She is also a trainer for training related to reliability engineering, asset life cycle analysis, and risk-based inspection, to name a few. Dr. Mokhtar is certified in API 580 Risk-Based Inspection.

Dr. Masdi bin Muhammad, CEng (UK) is an associate professor in the Department of Mechanical Engineering at Universiti Teknologi Petronas (UTP), Malaysia. He also leads the research group for Reliability Engineering and Asset Management at the university. Dr. Muhammad earned a BSc in Mechanical Engineering and an MSc in Manufacturing System Engineering from Lehigh University, Pennsylvania, USA. He earned his PhD in Mechanical Engineering at UTP with research on "Reliability Model for Repairable Systems with Multi-State Degradation." He is actively involved in research collaboration and consultancy with industries, and he supervises graduate students studying reliability and maintenance. He has twelve years of experience working in various positions at one of the leading semiconductor companies, involving process and equipment engineering, product development and material quality. He is a chartered engineer and member of the Board of Engineers Malaysia (BEM), the Institute of Engineers, Malaysia (IEM), and the American Society of Mechanical Engineers (ASME). He is also a Certified Reliability Engineer by the American Society for Quality (ASQ) and a Certified Maintenance and Reliability Professional by the Society for Maintenance and Reliability Professionals (SMRP).

Contents

Preface

Industrial engineering is an inter-disciplinary field designed to help organizations best utilize their resources in achieving their goals and objectives. This book presents the current state of the art of industrial engineering and provides useful information to those who wish to optimize their business practices. It will be of interest to researchers in the field of industrial engineering, industrial engineers and managers who want to develop systems to improve efficiency and productivity while increasing customer service and quality.

Dr. Ainul Akmar Mokhtar and Dr. Masdi Muhammad
Associate Professor,
Mechanical Engineering Department,
Universiti Teknologi Petronas,
Perak, Malaysia

Section 1

Optimization Methods in Plant Operation

Chapter 1

Supplier Evaluation and Selection in Automobile Industry

Lokpriya Gaikwad and Vivek Sunnapwar

Abstract

In automobile industry, to operate effectively the supply chain management, the purchasing function is very important to perform effectively. It is the responsibility of purchasing department to choose the correct suppliers to purchase the required products for their company. Thus, supplier evaluation technique is essential for purchase manager's point of view to choose the best supplier among available suppliers. The literature addresses quality, delivery, technology, value and service as the five most common criteria used for supplier quality evaluation. In this chapter, approach of evaluation and selection of supplier has been presented as per the ISO 9000/TS16949 standards. Considering the most important criteria for evaluating the quality of suppliers based on a review of the literature and observation in practice. Finally, these organizations continuously review and implement effective quality systems following the rigorous ISO 9000/TS16949 series of standards and most automobile companies have developed in-house procedures and software for the supplier selection process.

Keywords: supplier assessment, supplier selection and performance evaluation, supplier quality cost, supplier rating, part per million equivalents

1. Introduction

Conventionally firms have been divided in operational functions and each department take care of their own responsibility and manufacturing functions from procurement of raw material to dispatch of final products to the customer. Due to this reason, most of the organization purchasing commands a significant role, since purchased parts and components represent 40–60% of the sales [1] of its end products. This means with small cost saving in the acquisition of materials can have a greater impact on profits of the organization.

There has been an evolution in the role and structure of the purchasing function that gained great importance in the supply chain management due to the globalization and accelerated technological amend. It involves buying the raw materials and components for the organization to meet current need. The actions connected with it include selecting and qualifying suppliers, rating supplier performance, negotiating contracts, comparing price, quality and service, sourcing goods and service, timing purchases, selling terms of sale, evaluating the value received, predicting price, service, etc. Main responsibility of the purchasing department is the selection and evaluation of capable suppliers which brings financial benefits for the organization. The main objective of the supplier selection process is to reduce peril and maximize the total worth for the buyer organization considering strategic variables such as the choice between domestic and international suppliers, and the number of suppliers.

1.1 Literature reviews

Experts agree that no best way exists to evaluate and select suppliers, and thus organizations use a variety of approaches. The overall objective of the supplier evaluation process is to reduce risk and maximize overall value to the buyer. **Figure 1** presents the steps to follow when developing such a system [2].

Step 1: Identify key supplier evaluation categories

The first step in this process is identifying supplier evaluation criteria such as cost, quality, and on time delivery which are the important primary critical criteria that affect on the buyer. However, for critical items the supplier's in depth analysis related to their process capability or machine capability and ability to do a business is essential. For these reasons more supplier evaluation study is required. These criteria are typically the following:

A. Supplier managing capability

This is an essential way to assess, since management runs the business and makes the decisions that influence the future competitiveness of the vendor.

1. Overall workforce capabilities

This measurement requires an evaluation of third party personnel outside the organization. The reason is that well-known, self-motivated, stable employees should not be underestimated.

2. Cost composition

Accepting a supplier's total cost configuration helps a purchaser to determine how competently a supplier can produce things. A cost breakdown helps to identify probable areas of cost improvement.

3. Total quality management system

In supplier evaluation process, quality management systems at supplier end, their systems as well strategies must be address.

4. Technology and process capability, together with the supplier's design capability

This step helps to understand the technology, resource skill and capital requirement of the supplier during selection process.

5. Ecological regulation conformity

This is important given that purchasers do not want to be connected with be known ecological polluters from a public relations stand point.

6. Economic capability and steadiness

To check the economical capability of the supplier is essential for preliminary condition that the supplier must pass before a detailed evaluation can begin.

7. Production planning and control systems, including supplier on time delivery performance

The purpose behind this step is to evaluate the supplier from planning, scheduling and on time delivery point of view.

8. Information technology capability

Evidence that the supplier must use latest technology in their plant so that they can update their work environment.

9. Supplier purchasing strategies

These criteria are together one way to expand greater imminent and accepting of the supply chain of the suppliers.

10. Longer-term relationship probable

Supplier should be selected on the base of long term relationship, i.e., collaboration or partnership with the supplier. This will help both supplier and buyer to exchange ideas as well as technology with each other.

Step 2: Evaluation category weightage

The performance categories having weight reflect the relative importance of that category. The total of each weight must equal 1.0. That helps the management during the supplier selection and evaluation process.

Step 3: Identify and weight subcategories

In this, first identify performance subcategories within broader category in which the total sum of the subcategory weight must be equal to the total weight of the performance category.

Step 4: Identify scoring system for categories and subcategories

Scoring system takes criteria that may be highly skewed and develops a quantitative scale for measurement. Scoring system is effective if different individuals infer and score the same performance categories under assessment. For illustrative purposes, an example is a 5-point scale where 1 = poor, 2 = weak, 3 = marginal, 4 = qualified, and 5 = outstanding.

Step 5: Assess suppliers directly

A buyer can compare the scores of different suppliers for the same order and select one based on the evaluation score. It may be possible that supplier does not qualify at this time for further purchase consideration. Purchaser should have minimum acceptable performance necessities that suppliers must assure before they can become part of the supply base [3].

Step 6: Make selection based on evaluation results review

The major output from this step is a proposal about whether to accept a supplier for a business. A buyer may evaluate several suppliers who might be competing for a purchaser contract. The intention of the evaluation is to qualify potential suppliers for current or future business requirements.

Step 7: Review supplier performances constantly

After selecting a supplier, the supplier must perform as per buyer requirements to fulfill their needs. The prominence shifts from the initial evaluation and selection of suppliers to continuous improvement by suppliers into their process and product to fulfill buyer requirement.

Few authors have acknowledged criteria for supplier selection, such as the price, quality, and delivery, past supplier performance, capacity, information systems, service, and geographic location, among others [4–6]. These criteria are a key issue in the supplier measurement process since it dealings the performance of the suppliers.

1.2 Methodology

In most of the research which is based on supplier selection and evaluation, authors opined that the purchasing organizations use different approaches for evaluating and selecting supplier as per their requirements because of no best way is there.

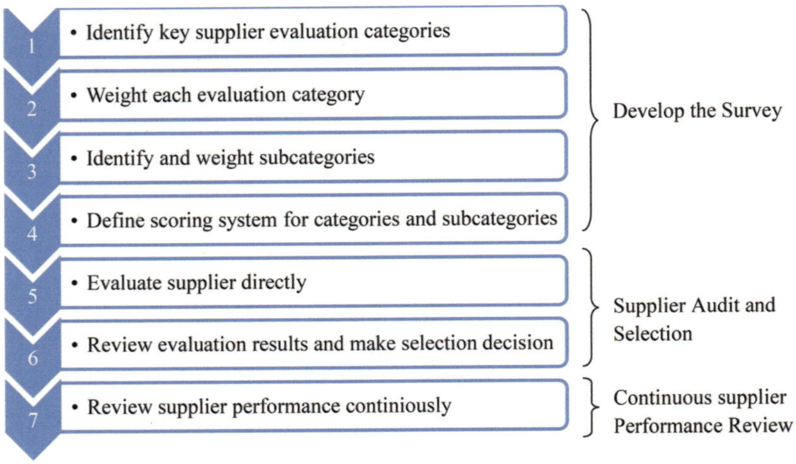

Figure 1.
Initial supplier evaluation and selection.

Supplier selection is based on a relative assessment using an exploratory case study approach which is generally used in most of the OEMs (Original Equipment Manufacturers).

2. Supplier evaluation and selection in auto industry

2.1 Rejected parts per million (PPM) level

PPMeq is a pointer for inward part quality by monitoring the performance of inward parts during on-going production, once parts is handed over by Component Development Materials Management (CDMM) to Supplier Quality Assurance (SQA) and procured by Supply Chain Management (SCM). This guideline sets the procedure of calculating the PPM Equivalent (PPMeq) and declaring non-conforming parts in the supplier deliveries. The suppliers' performance has assessed by the index PPM which has based only on rejection of parts at receiving stage and on line. But this index did not reflect the performance of the suppliers whose parts has mostly reworked on lines, the parts for which deviations/concessions were sought. Also much effort has been put in for segregating parts if any non-conformity has found in a lot. Some supplier parts also get rejected at the final assembly stage due to which the whole assembly faces rejection. To capture the effect of all the above conditions, a new index PPMeq has been formed.

Formula:

Supplier RPPM (rejected parts per million) is calculated on the basis of the amount of rejected parts versus the total amount of parts received in a given fiscal month. This computation is then normalized to replicate a continuous basis of one million units received.

- PPMeq can be calculated as:
  - Auto Sector-wise PPMeq
  - Plant-wise PPMeq
  - Supplier-wise PPMeq
  - Supplier Part-wise PPMeq

The agreement on PPM values does not signify a quality level accepted by Customer. All purchasing parts which are recognized as defective basically not be accepted and has been charged to the supplier.

PPMeq for a period is calculated as shown in Eq. (1)

$$PPMeq = \frac{N_{rej} + N_{rew} + N_{rej\,951} + 0.5 N_{dew} + 0.5 N_{conc} + 5.0 N_{rej\ at\ F.A.}}{N_{Total}} X\ 1000000 \quad (1)$$

where N_{rej} is the total inspection and line rejected quantity; N_{rew} is the total inspection and line quantity reworked; N_{dev} at F.A., is the total no. of quantity accepted under variation (inspection and line); N_{conc} is the total no. of quantity accepted under concession (inspection and line); N_{rej} at F.A., is the total no. of quantity rejected at Final Assembly; N_{Total} is the total no. of quantity received; and N_{rej} 951 is the total Quantity rejected on movement 951 for scrap at our end.

Example: A supplier ships 100,000 parts to a plant, of those 7 are found to be non-conforming.

The scorecard calculation will be (7/100,000) × 1,000,000 = 70 RPPM'S.

The Supplier's score for this example has 12 points.

Table 1 shows parts per million ranges and their respective scores.

From table it clear that, maximum score has been assigned to minimum rejected part per million and so on. Following are the minimum requirements from the suppliers end during the inspection of their submitted lot for acceptance to the OEMs.

Minimum expectations: it is expected that the minimum score should be 85%, (combined total of 51 points out of 60 possible).

Corrective actions: those suppliers who cannot meet the minimum expectation should be applied following corrective actions.

1. **First month:** announcement letters has been sent to Suppliers for giving justification regarding not meeting minimum score, reason for the same and what corrective action will be taken in future.

2. **Second consecutive month:** a second announcement letter has been sent stating that failing to meet minimum score and why. A corrective action plan is required.

3. **Third consecutive month:** if the problem is not solved in first and second notice then purchase manager either visited or called meeting with supplier to discuss performance. Within this period, suppliers may be on probation.

    - **Rejection:** any parts that not meeting customer specifications has rejected.

    - **Rework:** if doing any minor correction on part, it becomes fit for use then it comes under rework.

    - **Deviation:** any supplied part whose critical dimensions or material specifications cannot be reworked/repaired is termed as deviation.

    - **Concession:** when approval of product development has not taken for the minor repair that are not specified in the drawing and which does not affect the product quality, it is termed as concession.

    - **Segregation:** division of accepted/rejected parts, done with permission of QA personnel of that area, is known as segregation.

- **Rejection at final assembly:** due to supplier part if final assembly gets rejected, then it comes under rejection at final assembly.

To simplify communication and wherever technically practical and feasible, only one target value should be agreed for each product family delivered by suppliers or if possible for all products delivered.

General rules to declare Non-conforming parts:

1. **Rejection:** a part has been rejected if it falls out of the engineering specification. After proper inspection a part has been rejected. For example, in case of incorrect dimension, if the supplier has not produced a part as per specified dimensions, rejection will be done otherwise part will be accepted if it falls within specified specification.

2. **Rework:** parts shall be declared as rework if they are non-conforming only if the supplier is responsible. An analysis agreed by the supplier shall define the accountability for the rework (supplier); the supplier can take part in the investigation process. All records related to Deviation, Concession and Segregation, rework quantities are kept.

3. **Deviation:** the variation for the use of non-conforming part has been agreed by the Product development and respective plant quality head.

4. **Concession:** the dispensation has rose by the Manufacturing Quality and approved by the respective plant quality head.

5. **Segregation:** the quantities accepted after separation of the supplier parts has been taken into account in the calculation of PPMeq. This has not been interpreted as the total quantity which is separated.

6. **Quantity of non-conforming parts:** it has been declared with the conformity of the supplier and customer. Sample has not used to declare the non-conformance.

7. **Re-acceptance:** after correcting the parts it has been reaccepted within same month of rejection of that part so as to have correct performance of that supplier on monthly basis.

RPPM rating	Score	RPPM rating	Score
0–25	30	61–65	14
26–30	28	66–70	12
31–35	26	71–75	10
36–40	24	76–80	8
41–45	22	81–85	6
46–50	20	86–90	4
51–55	18	91–95	2
56–60	16	96–100	0

Table 1.
PPM ranges and their scores.

2.2 Scorecard

The purpose of the Scorecard is to communicate key supplier performance metrics that align with business objectives. The program rewards suppliers based on data, serves as a foundation for continuous improvement, and assists with future sourcing decisions.

Supplier rating system based on:

- **Delivery:** to meet expected dates as per delivery date.
- **Lead time:** time required in between placing order and receiving material.
- **Quality:** it is calculated in parts per million defective (PPM)
- **Productivity savings:** suppliers' assistance in helping to meet our productivity goals
- **Payment terms:** after delivery it may be within 3 to 6 month

Span:

To select a group of critical suppliers.

Benefits:

It benefits purchaser and supplier to achieve the benefits, sharing information by establishing open communication.

Supplier reimbursement:

- Visibly stated performance opportunity
- Enhance communication
- Business association get improved
- Data is available to measure performance
- Superior in general competitiveness in the market

Organization's reimbursement:

- Visibly communicated performance opportunity to supplier network
- Quicker associations with our suppliers
- Better accepting of overall performance

Supplier scorecard point system:

Every month suppliers receive performance score based on following areas as shown in **Table 2**.

Supplier performance levels:

Supplier is ranked depending upon their ongoing performance:

Level 1: supplier having 71 points or above is referred as world class and will be rewarded by new business opportunities.

Level 2: supplier having ongoing score 51–70 points performing acceptable level but at the same time assurance team should work to lift them to level 1 performance by developing them.

Level 3: a supplier having score 31–50 points kept on conditional level of performance. Assurance team must work to lift them to achieve level 2 or level 1 status.

Level 4: a supplier having score 30 points or below is a restricted supplier, any time they exist and another alternative source has to be find out.

2.3 Collaborate with suppliers

Enterprises that shared performance data with suppliers were able to generate 62% greater improvements in supplier performance than enterprises that only used this information internally (see **Figure 2**).

Based on a data purchaser can find out the opportunities to improve and develop the supplier performance by sharing business information. Many organizations can collaborate with their supplier in exchange of design and process data.

2.4 Process audit

Process audit for the manufacturing is quality tool to assess the continuous improvement of the organization in process/product performance. Doe to process audit, it become easy to analyze, maintain and improve quality system. So process audit become essential for the organization to sustain into competitive environment. Based on ISO 9001 certification requirement internal audit has to be conducted.

Sr. Nos.	Focus area	Allocated score
1.	Delivery	(0–20 points)
2.	Lead time	(0–20 points)
3.	Quality	(0–20 points)
4.	Productivity savings	(0–20 points)
5.	Payment terms	(0–20 points)
The maximum possible score for the Month or Year-to-Date is 100 points.		

Table 2.
Performance score based on the focus area.

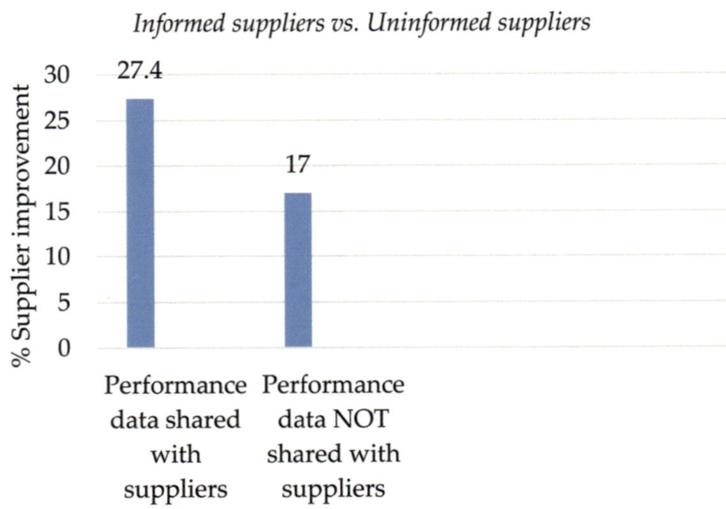

Figure 2.
Supplier performance improvements.

Due to customer awareness and global competition there is tremendous pressure on manufacturing firms to improve quality and reduce cost of the product. Each organization is struggling to meet customer varied demands as well as meet environmental regulation applied by government so that process/product should be ecological feasible. Audit is one of them to assess current state of nature and future plan of action to improve quality of the product and process. Audit can be used throughout the business area of quality, production, safety, human resource, purchasing, accounting, etc. [6]. Many organizations not only manufacturing but also service industry, have conducted the audit or been audited in order to comply with certain standard requirements [7].

2.5 ISO 9000 and ISO 14000 standards

The International Organization for Standardization (ISO) is a set of standards that "makes the development, manufacturing, and supply of products and services more efficient, safer and cleaner" [8]. The ISO has created several standards, but the best-known ones are ISO 9000 and ISO 14000. ISO 9000 is used for assessing quality requirements, while ISO 14000 is a standard for environmental quality management. ISO 9000 and ISO 14000 are known as "generic management system standards" because they can be applied to any product, material or service [8]. An ISO certificate can be given to any organization after it prepares its documents containing a description of its business practices in line with the guidelines provided by ISO. According to [9], having ISO 9000 has the following benefits:

1. **Way in to markets:** having ISO 9000 certification helps organizations to maintain and increase number of customers. Due to globalization, it is essential to certify and enter into global market.

2. **Customer requirement:** customers buy the product from certify supplier only. So now it become need that supplier should have ISO certification.

3. **Quality system improvement:** due to certification, quality system of the organization gets improved and also organization prepares itself for quality auditing.

4. **Other benefits:** the certificate is recognized around the world, and can develop quality through recovering an organization's overall competitiveness.

2.6 Normalize supplier measurement procedures across the enterprise

Organization should have formal measurement process which improves the supply base of the vendors than those vendors without having such type of formal measurement process.

On the other hand, firms should have standardized supplier performance metrics to reach better results at least 25% performance improvement can be seen in this process.

However, supplier performance metrics changes from organization to organization depending upon their needs and capabilities of the supplier supplying part or components to them. But agreeing on standard metrics for evaluating vendor performance is easy than on firm basis where different firm units have varied goals, requirements, and suppliers.

3. Results and discussion

A critical area to focus purchasing attention continues to be supplier quality management. Although supplier performance has likely improved in real terms over the last several years, supplier quality still does not fully satisfy continuously changing performance expectations. Measuring continuous supplier performance is not the only time when firms should evaluate suppliers. For most firms, supplier evaluation is central to their philosophy of quality at the source. Almost 70% of purchasing managers say that, the organizations use quantitative-based supplier evaluation process to check the capability and control techniques in there continuous quality improvement commitment. Around 80% firms assess supplier capabilities directly by cross-functional team site visits. Any kind of material or components either in semifinished, finished or raw material supplied by supplier has to be consider for calculating PPMeq and deciding which supplier has to be selected for doing business. If supplier having high PPM score consistently should be deleted from business list, PPM > 500 supplier has to be called or reviewed. Although most quality standards have been fulfilled till there are quality issues in manufacturing industry. According to ACMA reports 170 firms have already received ISO 9000 certification and 23 firms have received QS 9000 certification but still they are struggling to achieve excellence in quality.

In this work evaluation of the supplier has been done through Part per Million equivalents which help the OEMs to track the rejection rate at Plant level, Supplier level and Process level. Having high PPMeq the OEM can take a prompt decision regarding doing the business in future or to give time for improvement or to develop for excellence. Process-wise PPMeq can also help to find out the weak supplier for the particular process like Machining or Casting.

4. Conclusions

An organization should measure supplier's performance because without measurement improvement cannot become possible, also supplier cannot improve and remove wastages as well as cost drivers so vendors should be measured to facilitate performance improvement and enhance competitiveness. Thus an informed business decision has been possible that impact the enterprise.

Supplier quality cost should be incorporated into a buyer's supplier rating system. Supplier rating should be involved not only traditional indicators like cost, quality, on time delivery but also supplier quality costs. When problems that effect on customers occur, the speed with which problems has identified solved and the solution implemented has a direct impact on customer satisfaction. To reduce cost of quality there should be supplier involvement so that the quality related problems can be resolved and analyze the occurrence of the problems and failure that take place due to the faulty method or processes for that suppliers must be actively participated to solve the shop floor problems.

Use of part per million equivalent technique help to the purchasing organization to take a decision related to supplier selection and evaluation in critical conditions. It helps purchase manager in decision-making process at the time of selecting single supplier from the available number of suppliers.

Acknowledgements

The authors sincerely thank the editor and the anonymous reviewers for their constructive and important comments on the chapter.

Conflict of interest

I confirm there is no "conflict of interest."

Author details

Lokpriya Gaikwad[1*] and Vivek Sunnapwar[2]

1 Mechanical Engineering Department, SIES Graduate School of Technology, Navi Mumbai, India

2 Mechanical Engineering Department, Lokmanya Tilak College of Engineering, Navi Mumbai, India

*Address all correspondence to: lokpriya2007@gmail.com

References

[1] Ballou RH. Business Logistics Management. New Jersey: Prentice Hall; 1999

[2] Trent RJ, Handfield RB. Purchasing and Supply Chain Management. Cincinnati, Ohio: South-Western College Pub.; 2nd edition. 2002

[3] Dickson GW. An analysis of vendor selection systems and decisions. Journal of Purchasing. 1966;**1**:5-17

[4] Simpsom PM, Siguaw JA, White SC. Measuring the performance of suppliers: An analysis of evaluation processes. Journal of Supply Chain Management. 2006;**38**(4):29-41

[5] Dempsy WA. Vendor selection and the buying process. Industrial Marketing Management. 1978;7:257-267

[6] Weber CA, Current, Benton. Vendor selection criteria and methods. European Journal of Operational Research. 1991:2-18

[7] Askey JM, Dale BG. Internal quality management auditing: An examination. Managerial Auditing Journal. 1994;**9**(4):3-10

[8] Boyer K, Verma R. Operations and Supply chain Management for the 21st century. Mason: South-Western Cengage Learning; 2010

[9] Lio S, Wiebe H, Enke D. An expert advisory system for the ISO 9001 quality system. Experts Systems with Applications. 2004;**27**(2):313-322

Chapter 2

Optimal Control Promotional Policy for a New Product Incorporating Repeat Purchase in Segmented Market: A Control Theoretic Approach

Kuldeep Chaudhary and Prakash C. Jha

Abstract

This chapter considers an optimal control model to obtain dynamic promotional policies for a product considering a segmented market where first-time and additional repeat purchase sales are assumed to be generated through mass and differentiated promotions. Mass promotion is carried out in the whole market which reaches each segment with a fixed spectrum, and differentiated promotion is catered to each segment individually. The firm's finite promotional resources are to be allocated for promoting a product at mass and segment levels of the market in a finite time period. The formulated control problem obtains optimal promotional effort policy for each segment using the maximum principle. The applicability of the proposed control model is illustrated through a numerical example by discretizing the model.

Keywords: market segmentation, innovation-diffusion model, optimal control theory, maximum principle

1. Introduction

In the last few decades, the customers are made available with an increased amount of choices for particular goods or services. In such a situation to ensure that the customer chooses our product among others, it becomes important to communicate and inform about the innovative features and quality offered through the product and make a space in customers' minds. This task is achieved by promoting the product at regular intervals. Promotion plays a major role in raising customer awareness of the product, generates sales, and hence repeats purchases. Repeat purchase is an important phenomenon among the consumers that often measures their loyalty towards a brand. The higher is the repeat purchase value, it can be said that the better a firm is doing to keep customers loyal. This chapter focuses on determining the optimal promotional effort policies for a consumer durable product by assuming that the single purchase and the repeat purchase of a product are generated through the combined effect of mass and differentiated promotions in a segmented market.

IntechOpen

Promotional strategies are often targeted to a potential market chosen in accordance with the firm's product type. Once target market is decided, market segmentation is carried out to divide the broad target market into subsets of consumers who have common needs and priorities, and then designing and implementing strategies are done to target them. Market segmentation plays an important role in development of the marketing strategies. Different customers have different needs, and it is impossible to satisfy all customers treating them alike. Promotional policies for the products are built by considering the heterogeneity in the potential market. Firms that identify the specific needs of the groups of customers are able to develop the right offer for the submarkets and obtain a competitive advantage over other firms. The concept of market segmentation emerged, as the market-oriented thought evolved among the firms. Market segmentation has thus become the building block of the effective promotional planning. It partitions the markets into groups of potential customers on the basis of geographic, demographic, and psychographic variables and behavioural customer characteristics.

Once the segmentation process is complete, the next step following it is choosing the targeting strategies that can be implemented. The firm must decide whether they want to choose segment-specific or mass (differentiated) promotional strategies. Mass promotion is implemented by treating the market as homogeneous and giving common message in all the segments through mass communication, the effect of which reaches each of the segments proportionally known as spectrum effect. However, the preferences of customers may differ, and same offering may not affect all potential customers and urge them towards product adoption. If firms ignore these differences, another competing firm can market similar product serving specific groups, and this may lead to losing customers. Segment-specific promotion recognizes this diversified customer base and takes into consideration the varying consumers in different segments. The promotional messages are constructed accordingly here. Both the mass- and segment-specific strategies play important roles and have their own advantages. Firms generally promote their product in the market at both the levels mass and segment. In this chapter, we assume that the evolution of sales of the product is through mass and differentiated promotions and build a control model for determining the promotional policies that maximizes the total profit constrained on the total budget. The promotion effort policies are generated by using the maximum principle. The model proposed is continuous in nature, but in practical the data available is discrete. Also the model is nonlinear and becomes NP-hard in nature. Thus we have used Lingo11 to solve the discretized version and show the model application.

The rest of this chapter is organized as follows. In Section 2 of this chapter, we provide a brief literature review and in Section 3, we introduce the diffusion model with repeat purchasing and discuss its optimal control formulation and develop segmented sales rate under the assumption that the practitioner may choose independently the advertising intensity directed towards each segment as well as combined advertising intensity. The problem is discussed, and it is solved using Pontryagin's maximum principle with particular cases in Section 4. Section 5 gives the numerical illustration for the discretized version of the problem using Lingo11 software and finally in Section 6, we conclude our chapter.

2. Literature review

Few people have worked in optimal control theory considering market segmentation in advertising models [1–3]. A discrete time stochastic model of multiple media selection in a segmented market was analysed by Little and Lodish [1].

Seidmann et al. [2] proposed a general sales-advertising model in which the state of the system represents a population distribution over a parameter space, and they show that such models are well posed and that there exists an optimal control. Buratto et al. [4] have given some market segmentation concepts into advertising models during the introduction of new product and advertising processes for sales over an infinite horizon. Grosset and Viscolani [3] discussed the optimal advertising policy for a new product introduction considering only the external influence in a segmented market with Nerlove-Arrow's [5] linear goodwill dynamics. Nerlove and Arrow [5] proposed a model in which the effect of advertising on sales is mediated by the goodwill variable. The goodwill state variable represents the effects of the firm investment in advertising, and it affects the demand of the product together with price and other external factors. From past few years, a number of researchers have been working in the area of optimal control models pertaining to advertising expenditure and price in marketing [6]. The simplest diffusion model was due to Bass [7]. Since the landmark work of Bass, the model has been widely used in the diffusion theory. The major limitation of this model is that it does not take into consideration the impact of marketing variables. Many authors have suitably modified the Bass model to study the impact of price on new product diffusion [8–13]. These models incorporate the pricing effects on diffusion. Also there are models that incorporate the effect of advertising on diffusion [9, 14, 15]. Horsky and Simmon [9] incorporated the effects of advertising in the Bass innovation coefficient. Thompson and Teng [16] incorporated learning curve production cost in their oligopoly price-advertising model. Bass et al. [17] included both price and advertising in their generalized Bass model.

Jha et al. [18] used the concept of market segmentation in diffusion model for advertising a new product and studied the optimal advertising effectiveness rate in a segmented market. They discussed the evolution of sales dynamics in the segmented market under two cases. Firstly, they assumed that the firm advertises in each segment independently, and further they took the case of a single advertising channel, which reaches several segments with a fixed spectrum. Manik et al. [19] amalgamated the two problems formulated by Jha et al. [18] and formulated an optimal control problem where they studied the effect of differentiated promotional effort and mass promotional effort on evolution of sales rate for each segment. They obtained the optimal promotional effort policy for the proposed model. Dynamic behaviour of optimal control theory leads to its application in sales-promotion control analysis and provides a powerful tool for understanding the behaviour of sales-promotion system where dynamic aspect plays an important role. Numerous papers on the application of optimal control theory in sales-advertising problem exist in the literature [20, 21]. However the literature missed out the control model to determine the control policies in a segmented market considering repeat purchasers in the sales through mass and differentiated promotions and taking the budget constraint which we try to do in this chapter.

3. Model development

We begin our analysis by stating the following assumption that $M(>1)$ is the total market segments and a discrete variable. The sum $\sum_{i=1}^{M}\overline{X_i}$ denotes the total number of potential customers of the product in all the segments. The firm simultaneously uses mass market promotion and differentiated market promotion to capture the potential market in each segment, respectively. Mass market promotion reaches each segment proportionally called segment-specific spectrum. Let $x_i(t)$ be the number of adopter by time t for the ith segment. During diffusion process, repeat purchases of the product may also occur, and those adopters who have already adopted may repurchase the product again. Therefore, the number of adopters for a new product can increase due to both first purchase and repeat purchasing. Under the influence of mass market and differentiated market promotion, evolution of sales rate [7] can be described by the following differential equation:

$$\frac{dx_i(t)}{dt} = b_i(t)(u_i(t) + \alpha_i u(t))\left(\overline{X_i} - (1 - g_i)x_i(t)\right), i = 1, 2, \ldots, M \tag{1}$$

with the initial condition $x_i(0) = x_{i0} \forall i = 1, 2, \ldots, M,$ where α_i denotes the segment spectrum of mass promotion $\left(\alpha_i > 0 \ \& \ \sum_{i=1}^{M} \alpha_i = 1\right)$; $g_i \left(0 \le g_i \le 1\right)$ is susceptible to repeat purchasing, and repeat purchasing is influenced by all factors (both internal and external) affecting first purchase in i^{th} segment by time t; $u_i(t)$ is differentiated promotional effort rate for i^{th} segment at time t; *and* $u(t)$ is mass market promotional effort rate at time t, and $b_i(t)$ is the adoption rate per additional adoption for the ith segment. $b_i(t)$ can be represented either as a function of time or as a function of the number of previous adopters. Since the latter approach is used most widely, it is the one applied here. Therefore, Eq. (1) can be rewritten as follows:

$$\frac{dx_i(t)}{dt} = \left(p_i + q_i \frac{x_i(t)}{\overline{X_i}}\right)(u_i(t) + \alpha_i\, u(t))\left(\overline{X_i} - (1 - g_i)x_i(t)\right), i = 1, 2, \ldots, M \tag{2}$$

where p_{i} and q_{i} are coefficients of external and internal influences in i^{th} segment, respectively.

The objective of the firm is to maximize the present value of the profit in a planning horizon for a segmented market by selecting optimal mass and differentiated promotional effort rates for the firm. Thus, the objective function can be represented by

$$\text{Max}\, J = \int_0^T e^{-\gamma t}\left(\sum_{i=1}^{M} [(P_i - C_i(x_i(t))\dot{x}(t) - \phi_i(u_i(t))] - \varphi(u(t))\right) dt \tag{3}$$

where $\phi_i(u_i(t))$ and $\varphi(u(t))$ are differentiated market promotional effort and mass market promotional effort cost, respectively, γ is discounted profit, P_i is sales price for i^{th} segment, and $C_i(x_i(t))$ is production cost per unit for i^{th} segment, that is, continuous and differentiable with assumption $C_i'(.) > 0$ and $P_i - C_i(x_i(t)) > 0$.

During the promotion, differentiated and mass promotions are competing for the limited promotion budget expenditure. Therefore, firms monitor the promotion strategy in all segments closely and allocate their promotional expenditure budget optimally among these segments. The budget constraint for all segments is represented as

$$\int_0^T \left(\sum_{i=1}^{M} \phi_i(u_i(t)) + \varphi(u(t))\right) dt \le W_0 \tag{4}$$

where W_0 is the fixed budget expenditure for all segments over time. Constraint (4) corresponds to the common promotional expenditure capacity that is allocated among all the segments. This constraint couples the segment and prevents us from simply solving M times a single-segment problem. The above problem can be written as an optimal control problem:

$$\left.\begin{aligned} \text{Max}\,J &= \int_0^T e^{-rt}\left(\sum_{i=1}^M [(P_i - C_i(x_i(t))\dot{x}(t) - \phi_i(u_i(t))] - \varphi(u(t))\right) dt \\ \frac{dx_i(t)}{dt} &= \left(p_i + q_i \frac{x_i(t)}{\overline{X_i}}\right)(u_i(t) + \alpha_i\, u(t))\left(\overline{X_i} - (1-g_i)x_i(t)\right), i = 1, 2, \dots, M \\ & x_i(0) = x_{i0} \forall i = 1, 2, \dots M \\ & \int_0^T \left(\sum_{i=1}^M \phi_i(u_i(t)) + \varphi(u(t))\right) dt \le W_0 \end{aligned}\right\} \quad (5)$$

The above formulated optimal control problem consists of $2M + 1$ control variables $(u_i(t), u(t))$ and M state variables $(x_i(t))$.

4. Solution approach

To solve the above optimal control theory problem, we define a new state variable $W(t) = W_0 - \int_0^t \left(\sum_{i=1}^M \phi_i(u_i(t)) + \varphi(u(t))\right) dt$ with $W(0) = W_0$ and $W(T) \ge 0$. With new state variable, we rewrite the above optimal control problem (5) as

$$\left.\begin{aligned} \text{Max}\,J &= \int_0^T e^{-rt}\left(\sum_{i=1}^M [\,(P_i - C_i(x_i(t)))\dot{x}(t) - \phi_i(u_i(t))] - \varphi(u(t))\right) dt \\ \frac{dx_i(t)}{dt} &= \left(p_i + q_i \frac{x_i(t)}{\overline{X_i}}\right)(u_i(t) + \alpha_i\, u(t))\left(\overline{X_i} - (1-g_i)x_i(t)\right), i = 1, 2, \dots, M \\ & x_i(0) = x_{i0} \forall i = 1, 2, \dots M \\ \dot{W}(t) &= -\left(\sum_{i=1}^M \phi_i(u_i(t)) + \varphi(u(t))\right), W(0) = W_0, W(T) \ge 0 \end{aligned}\right\} \quad (6)$$

Now, we obtain an optimal control problem with $2M + 1$ control variable and $M + 1$ state variable for all segments. Using the maximum principle [22], Hamiltonian can be defined as

$$H = \begin{pmatrix} \sum_{i=1}^M [\,(P_i - C_i(x_i(t)) + \lambda_i(t))\,\dot{x}(t) - \phi_i(u_i(t))] - \varphi(u(t)) \\ -\mu(t)\left(\sum_{i=1}^M \phi_i(u_i(t)) + \varphi(u(t))\right) \end{pmatrix} \quad (7)$$

The Hamiltonian represents the overall profit of the various policy decisions with both the immediate and the future effects taken into account. Assuming the existence of an optimal control solution, the maximum principle provides the necessary optimality conditions; there exist piecewise continuously differentiable functions $\lambda_i(t)$ and $\mu(t)$ for all $t \in [0, T]$. The value of $\lambda_i(t)$ and $\mu(t)$ define marginal valuation of state variables $x_i(t)$ and $W(t)$ at time t, respectively. Here, $\lambda_i(t)$ stands for change in future profit as making a small in $x_i(t)$ at time t, and $\mu(t)$ is the future profit of promotional effort per unit promotion effort expenditure at time t. These variables are known as adjoint variables and describe the similar behaviour in optimal control theory as dual variables in nonlinear programming.

From the necessary optimality conditions [22, 23] of maximum principle, we have

$$H\left(t, x_i^*, u_i^*, u^*, \lambda, \mu\right) = H\left(t, x_i^*, u_i, u, \lambda, \mu\right) \quad (8)$$

$$\frac{\partial H^*}{\partial u_i} = 0 \tag{9}$$

$$\frac{\partial H^*}{\partial u} = 0 \tag{10}$$

$$\frac{d\lambda_i(t)}{dt} = \gamma\,\lambda_i(t) - \frac{\partial H^*}{\partial x_i(t)}, \lambda_i(T) = 0 \tag{11}$$

$$\frac{d\mu(t)}{dt} = \lambda\,\mu(t) - \frac{\partial H^*}{\partial W(t)}, \mu(T) \geq 0, \tag{12}$$

$$W(T) + W_0 \geq 0, \mu(T)(W(T) + W_0) = 0 \tag{13}$$

Here, $\mu(T) \geq 0$, $W(T) + W_0 \geq 0$, $\mu(T)(W(T) + W_0) = 0$ are called as transversality conditions for $W(t)$. Here, Hamiltonian is independent to $W(t)$, and then we have $\dot{\mu} = \gamma\mu - \frac{\partial H}{\partial W} \Longrightarrow \mu(t) = \mu_T\, e^{\gamma(t-T)}$. Hence, it is clear that the multiplier associated with any integral constraint is constant over time irrespective of their nature (i.e. whether equality or inequality). The Hamiltonian H of each of the segments is strictly concave in $u_i(t)$ and $u(t)$. According to the Mangasarian sufficiency theorem [22, 23], there exist unique values of promotional effort controls $u_i^*(t)$ and $u^*(t)$ for each segment, respectively. From Eqs. (9) and (10), we get

$$u_i^*(t) = \phi_i^{-1}\left(\frac{(P_i - C_i(x_i(t)) + \lambda_i(t))\frac{\partial \dot{x}_i(t)}{\partial u_i} - \frac{\partial C_i}{\partial x_i}\frac{\partial x_i}{\partial u_i}\dot{x}_i}{1 + \mu_T\, e^{\gamma(t-T)}}\right), i = 1, 2, \ldots, M \tag{14}$$

$$u(t) = \varphi^{-1}\left(\frac{\sum_{i=1}^{M}\left((P_i - C_i(x_i(t)) + \lambda_i(t))\frac{\partial \dot{x}_i(t)}{\partial u_i}\right) - \frac{\partial C_i}{\partial x_i}\frac{\partial x_i}{\partial u_i}\dot{x}_i}{1 + \mu_T\, e^{\gamma(t-T)}}\right), i = 1, 2, \ldots, M \tag{15}$$

where ϕ_i^{-1} and φ^{-1} are inverse functions of ϕ_i and φ, respectively. If we assume product cost is independent to $x_i(t)$, i.e. $C_i(x_i(t)) = C_i$, then optimal promotional effort policies for each segment become

$$u_i^*(t) = \phi_i^{-1}\left(\frac{(P_i - C_i + \lambda_i(t))\left(p_i + q_i\frac{x_i(t)}{\overline{X}_i}\right)\left(\overline{X}_i - (1 - g_i)x_i(t)\right)}{1 + \mu_T\, e^{\gamma(t-T)}}\right), i = 1, 2, \ldots, M \tag{16}$$

$$u(t) = \varphi^{-1}\left(\frac{\sum_{i=1}^{M}\left((P_i - C_i + \lambda_i(t))\,\alpha_i\left(p_i + q_i\frac{x_i(t)}{\overline{X}_i}\right)\left(\overline{X}_i - (1 - g_i)x_i(t)\right)\right)}{1 + \mu_T\, e^{\gamma(t-T)}}\right), i = 1, 2, \ldots, M \tag{17}$$

The optimal control promotional policy shows that when market is almost saturated, then differentiated market promotional expenditure rate and mass market promotional expenditure rate, respectively, should be zero (i.e. there is no need of promotion in the market).

For optimal control policy, the optimal sales trajectory using optimal values of differentiated market promotional effort $\left(u_i^*(t)\right)$ and mass market promotional effort $\left(u^*(t)\right)$ rates for each segment are given by

$$x_i^*(t) = \frac{\overline{X_i}\left(\left(\frac{p_i + q_i \frac{x_i(0)}{\overline{X_i}}}{\overline{X_i} - (1-g_i)x_i(0)}\right) \exp\left((q_i + p_i(1-g_i)) \int_0^t (u_i^*(\tau) + \alpha_i u^*(\tau))d\tau\right)\right) - p_i}{\frac{q_i}{\overline{X_i}} + (1-g_i)\left(\frac{p_i + q_i \frac{x_i(0)}{\overline{X_i}}}{\overline{X_i} - (1-g_i)x_i(0)}\right) \exp\left((q_i + p_i(1-g_i)) \int_0^t (u_i^*(\tau) + \alpha_i u^*(\tau))d\tau\right)} \quad (18)$$

If $x_i(0) = 0$, then we get the following result:

$$x_i^*(t) = \frac{1 - \exp\left(-(q_i + p_i(1-g_i)) \int_0^t (u_i^*(\tau) + \alpha_i u^*(\tau))d\tau\right)}{(1-g_i) + \frac{q_i}{p_i} \exp\left(-(q_i + p_i(1-g_i)) \int_0^t (u_i^*(\tau) + \alpha_i u^*(\tau))d\tau\right)}, i = 1, 2, \ldots, M \quad (19)$$

and adjoint trajectory is given as

$$\frac{d\lambda_i(t)}{dt} = \gamma\, \lambda_i(t) - \left\{ (P_i - C_i(x_i(t)) + \lambda_i(t))\left(\frac{\partial \dot{x}_i}{\partial x_i}\right) - \dot{x}_i(t)\left(\frac{\partial C_i(x_i(t))}{\partial x_i(t)}\right) \right\} \quad (20)$$

with transversality condition $\lambda_i(T) = 0$. Integrating (20), the value of future profit of having one more unit of sales is

$$\lambda_i(t) = e^{-\gamma t} \int_t^T e^{-\gamma s} \left((P_i - C_i + \lambda_i(t))\left(\frac{\partial \dot{x}_i}{\partial x_i}\right) - \dot{x}_i(t)\left(\frac{\partial C_i}{\partial x_i}\right) \right) dt \quad (21)$$

4.1 Particular cases

4.1.1 When differentiated market promotional effort and mass market promotional effort costs are linear functions

Let us assume that differentiated market promotional effort and mass market promotional effort costs take the following linear forms: $\phi_i(u_i(t)) = \kappa_i\, u_i(t)$ and $\varphi(u(t)) = \kappa\, u(t)$ and $\overline{a}_i \le u_i(t) \le \overline{A}_i$, $\overline{a} \le u(t) \le \overline{A}$, where $\overline{a}_i, \overline{A}_i, \overline{a}$, and $\overline{A}$ are positive constants which are minimum and maximum acceptable promotional effort rates ($\overline{a}_i, \overline{A}_i, \overline{a},$ and $\overline{A}$ are determined by the promotional budget) and κ_i is the per unit cost of promotional effort per unit time towards i^{th} segment and κ is the per unit cost of promotional effort per unit time towards mass market. Now, Hamiltonian can be defined as

$$H = \begin{pmatrix} \sum_{i=1}^{M} [(P_i - C_i(x_i(t)) + \lambda_i(t))\, \dot{x}_i(t) - \kappa_i\, u_i(t)] - \kappa\, u(t) \\ -\mu(t)\left(\sum_{i=1}^{M} \kappa_i\, u_i(t) + \kappa\, u(t) \right) \end{pmatrix} \quad (22)$$

Since Hamiltonian is linear in $u_i(t)$ and $u(t)$, optimal differentiated market promotional effort and mass market promotional effort as obtained by the maximum principle are given by

$$u_i^*(t) = \begin{cases} \overline{a}_i \text{ if } B_i \le 0 \\ \overline{A}_i \text{ if } B_i > 0 \end{cases} \quad (23)$$

$$u^*(t) = \begin{cases} \overline{a} \text{ if } D \le 0 \\ \overline{A} \text{ if } D > 0 \end{cases} \quad (24)$$

where $B_i = (P_i - C_i + \lambda_i(t))\left(p_i + q_i \frac{x_i}{\overline{X}_i}\right)(\overline{X}_i - (1 - g_i)x_i(t)) - \kappa_i(1 + \mu(t))$ and $D = \sum_{i=1}^{M}\left(\alpha_i(P_i - C_i + \lambda_i(t))\left(p_i + q_i \frac{x_i}{\overline{X}_i}\right)(\overline{X}_i - (1 - g_i)x_i(t))\right) - \varepsilon(1 + \mu(t))$ are promotional effort switching functions and called 'bang-bang' control. However, interior control is possible on an arc along $u_i(t)$ and $u(t)$. Such an arc is known as the 'singular arc' [22].

This optimal control advertising policy shows that when market is almost saturated, then our switching

functions $B_i = (P_i - C_i + \lambda_i(t))\left(p_i + q_i \frac{x_i}{\overline{X}_i}\right)(\overline{X}_i - (1 - g_i)x_i(t)) - \kappa_i(1 + \mu(t))$ and $D = \sum_{i=1}^{M}\left(\alpha_i(P_i - C_i + \lambda_i(t))\left(p_i + q_i \frac{x_i}{\overline{X}_i}\right)(\overline{X}_i - (1 - g_i)x_i(t))\right) - \varepsilon(1 + \mu(t))$ become negative or zero. Therefore, optimal advertising policy shows that there is no need to spend money, time, or resources on advertising, i.e. we do the advertising with minimum effectiveness rate.

There are four possible sets of optimal control values of differentiated market promotional effort $(u_i^*(t))$ and mass market promotional effort $(u^*(t))$ rate (**Figures 1** and 2): (1) $u_i^*(t) = \overline{a}_i, u^*(t) = \overline{a},$ (2)

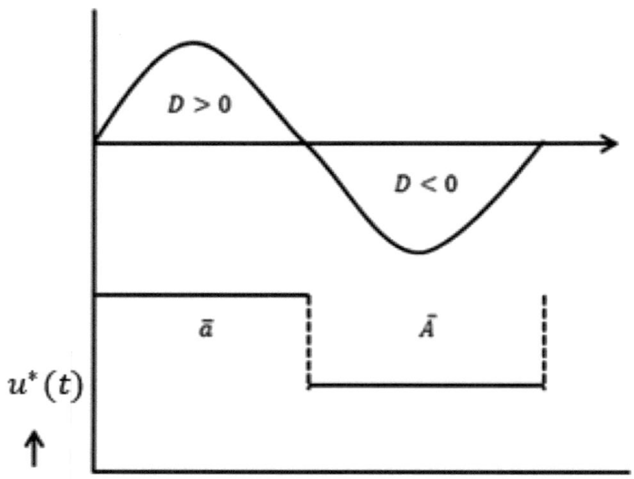

Figure 1.
Optimal promotional effort allocation policy for mass market promotional effort.

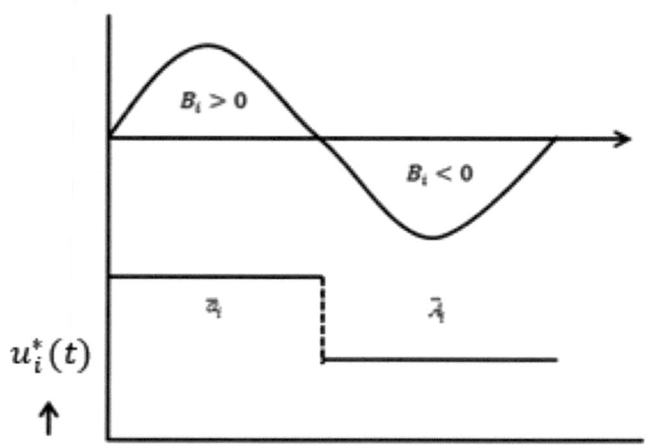

Figure 2.
Optimal promotional effort allocation policy for differentiated market promotional.

$u_i^*(t) = \overline{a}_i, u^*(t) = \overline{A}$, (3) $u_i^*(t) = \overline{A}_i, u^*(t) = \overline{a}$, and (4) $u_i^*(t) = \overline{A}_i, u^*(t) = \overline{A}$. Using these optimal values of differentiated market promotional effort $(u_i^*(t))$ and mass market promotional effort $(u^*(t))$ for each segment, we can obtain the optimal sales trajectories and adjoint trajectories. If we consider optimal values $u_i^*(t) = \overline{A}_i, u^*(t) = \overline{A}$, then the optimal sales and adjoint values can be described as

$$x_i^*(t) = \frac{\overline{X}_i\left(\left(\frac{p_i + q_i\frac{x_i(0)}{\overline{X}_i}}{\overline{X}_i - (1-g_i)x_i(0)}\right)\exp\left((q_i + p_i(1-g_i))(\overline{A}_i + \alpha_i\overline{A}_i)t\right)\right) - p_i}{\frac{q_i}{\overline{X}_i} + (1-g_i)\left(\frac{p_i + q_i\frac{x_i(0)}{\overline{X}_i}}{\overline{X}_i - (1-g_i)x_i(0)}\right)\exp\left((q_i + p_i(1-g_i))(\overline{A}_i + \alpha_i\overline{A}_i)t\right)} \forall i = 1, 2, 3, \ldots, M \tag{25}$$

If $x_i(0) = 0$, then we get the following result

$$x_i^*(t) = \frac{1 - \exp\left(-(q_i + p_i(1-g_i))(\overline{A}_i + \alpha_i\overline{A}_i)t\right)}{(1-g_i) + \frac{q_i}{p_i}\exp\left(-(q_i + p_i(1-g_i))(\overline{A}_i + \alpha_i\overline{A}_i)t\right)}, i = 1, 2, \ldots, M \tag{26}$$

which is similar to Bass model [7] sales trajectory with repeat purchasing, and the adjoint variable is given by

$$\frac{d\lambda_i(t)}{dt} = \rho\lambda_i(t) - \left((P_i - C_i + \lambda_i)(\overline{A}_i + \alpha_i\overline{A})((1-g_i)(2x_i^* - X_i) - X_i)\right), \lambda_i(T) = 0 \tag{27}$$

The value of $\lambda_i(t)$ stands for per unit change in future profit of having one more unit of variable $x_i(t)$.

4.1.2 When differentiated market promotional effort and mass market promotional effort costs are quadratic functions

Promotional efforts towards differentiated market and mass market are costly. Let us assume that differentiated market promotional effort and mass market promotional effort costs take the following quadratic forms $\phi_i(u_i(t)) = \frac{\kappa_i}{2}u_i^2(t)$ and $\varphi(u(t)) = \frac{\kappa}{2}u^2(t)$ where $\kappa_i > 0$ and $\kappa > 0$ are positive constants and represent the magnitude of promotional effort rate per unit time towards i^{th} segment and towards mass market, respectively. This assumption is common in literature [24], where promotion cost is quadratic. Now, Hamiltonian can be defined as

$$H = \begin{pmatrix} \sum_{i=1}^{M}\left[(P_i - C_i(x_i(t)) + \lambda_i(t))\dot{x}(t) - \frac{\kappa_i}{2}u_i^2(t)\right] - \frac{\kappa}{2}u^2(t) \\ -\mu(t)\left(\sum_{i=1}^{M}\frac{\kappa_i}{2}u_i^2(t) + \frac{\kappa}{2}u^2(t)\right) \end{pmatrix} \tag{28}$$

From the optimality necessary conditions (6), the optimal differentiated market promotional effort and mass market promotional effort are given by

$$u_i^*(t) = \frac{1}{\kappa_i}\left(\frac{(P_i - C_i + \lambda_i(t))\left(p_i + q_i\frac{x_i(t)}{\overline{X}_i}\right)(\overline{X}_i - (1-g_i)x_i(t))}{1 + \mu_T e^{\gamma(t-T)}}\right) \tag{29}$$

$$u(t) = \frac{1}{\kappa}\left(\frac{\sum_{i=1}^{M}\left(\left(P_i - C_i + \lambda_i(t)\right)\alpha_i\left(p_i + q_i\frac{x_i(t)}{\overline{X_i}}\right)\left(\overline{X_i} - (1-g_i)x_i(t)\right)\right)}{1 + \mu_T\, e^{\gamma(t-T)}}\right) \tag{30}$$

Using optimal differentiated market promotional effort and mass market promotional effort rates from above Eqs. (29) and (30), we can obtain the optimal sales trajectories. Due to cumbersome analytical expression and an aim to illustrate the applicability of the formulated model through a numerical example, the discounted continuous optimal problem (5) is transformed into equivalent discrete problem [25] which can be solved using differential evolution. The equivalent discrete optimal control of the budgetary problem can be written as follows:

$$\left.\begin{aligned} &\operatorname{Max} J = \sum_{k=1}^{T}\left(\left(\begin{matrix}\left[\sum_{i=1}^{M}(P_i - C_i(k))(x_i(k+1) - x_i(k) - \phi_i(u_i(k)))\right] \\ -\varphi(u(k))\end{matrix}\right)\left(\frac{1}{(1+\gamma)^{k-1}}\right)\right) \\ &\qquad\qquad\qquad\qquad\qquad\qquad\qquad\qquad\textit{subjected to} \\ &x_i(k+1) = x_i(k) + \left(p_i + q_i\frac{x_i(k)}{\overline{X_i}}\right)(u_i(k) + \alpha_i\, u(k))\left(\overline{X_i} - (1-g_i)x_i(k)\right), i = 1, 2, \dots, M \\ &\qquad\qquad\qquad\sum_{k=1}^{T}\left(\sum_{i=1}^{M}(\phi_i(u_i(k))) + \varphi(u(k))\right) \le W_0 \end{aligned}\right\} \tag{31}$$

The discretized version of the model is NP-hard; therefore, we use Lingo11 [26] to solve the discrete formulation.

5. Numerical illustration

To validate the model formulation, we consider a case of a company that has to find the optimal advertising policies for its consumer durable product. The company advertises at both national and regional levels of the market. To find the advertising policy for four segments, the values of the parameters, price, and cost of the product are given in **Table 1**.

	S1	S2	S3	S4
$\overline{N}_i$	279106.6	152460.1	97580.78	215868.5
p_i	0.000766	0.001161	0.00138	0.000549
q_i	0.137605	0.480576	0.540395	0.31362
α_i	0.3	0.19	0.189	0.320568
g_i	0.05	0.0265	0.0878	0.047644
κ_i (in ₹)	243,961	388,753	336,791	517,530
ε(in ₹)	1,153,922			
P_i	400,000	440,000	420,000	450,000
C_i	340,000	370,000	340,000	390,000
Initial sales$_i$	8969	8000	8000	8000

Table 1.
Parameters.

The discrete optimal control problem developed in this chapter is solved using differential evolution. Total promotional budget is assumed to be ₹ 3,000,000,000 which has to be allocated for mass market promotion and segment-specific promotion in four segments of the market. The time horizon has been divided into 12 equal time periods. The number of market segments is four (i.e. $M = 4$). The problem is coded in Lingo11 and solved.

Optimal allocation of promotional effort resources by solving each segment is given in **Table 2** for both mass and differentiated promotions, and the corresponding sales is tabulated in **Table 3**.

	Differentiated				Mass
	S1	S2	S3	S4	
T1	13.61	2.14	5.16	1.00	12.62
T2	14.69	1.30	2.42	1.00	13.20
T3	14.71	1.57	5.61	1.09	14.13
T4	17.06	6.87	1.56	1.50	15.02
T5	7.56	2.07	6.03	1.00	15.91
T6	19.63	2.32	1.96	1.00	16.80
T7	10.02	2.56	2.16	1.00	17.67
T8	21.99	4.88	6.62	3.06	16.61
T9	11.17	3.01	2.54	1.32	19.35
T10	24.28	8.27	7.00	2.74	20.19
T11	13.54	3.43	2.90	3.11	21.01
T12	26.44	8.68	7.34	10.74	28.78

Table 2.
Optimal differentiated and mass promotional allocations (in units).

	S1	S2	S3	S4
T1	8969	8000	8000	8000
T2	33,386	25,338	39,111	20,804
T3	112,850	64,724	105,406	52,397
T4	292,429	142,818	112,330	124,026
T5	296,492	201,723	98,941	235,744
T6	291,852	59,131	135,301	217,610
T7	297,762	157,208	35,605	235,055
T8	289,245	155,499	106,680	216,804
T9	305,967	158,657	108,227	241,642
T10	276,207	151,979	103,967	203,831
T11	345,626	178,184	121,081	263,316
T12	178,359	90,396	61,631	131,523

Table 3.
Optimal sales from potential market.

In the above case, we have solved the discretized problem by taking differentiated and mass promotional efforts as a linear function.

6. Conclusion

This chapter formulates an optimal control problem to find the optimal promotional policies for a consumer durable product in a segmented market where the sales are evolved through the combination of two promotion strategies: mass and differentiated promotions. The sales include the first-time purchase and the repeat purchases built through loyalty towards the product. Also to make the problem more realistic, we take a total budget constraint. The objective is to maximize the total profit through promotion. Maximum principle has been used to obtain the solution of the proposed problem. After discretizing the problem with linear costs, a numerical example has been solved using Lingo11 to illustrate the applicability of the approach. The developed optimal control model can be further extended in several ways. For instance, factors such as price, quality, and cost can be incorporated along with differentiated and mass market promotional effort expenditures. Further this monopolistic model can also be extended to competitive duopolistic or oligopolistic markets. Also the model can be extended to obtain optimal control policies for two and/or more generations' products in the market.

Conflict of interest

The authors declare that there is no conflict of interests regarding the publication of this chapter.

Author details

Kuldeep Chaudhary[1*] and Prakash C. Jha[2]

1 Department of Applied Mathematics, Amity Institute of Applied Sciences, Amity University, Noida, India

2 Department of Operations Research, University of Delhi, Delhi, India

*Address all correspondence to: kchaudhary26@amity.edu

References

[1] Little JDC, Lodish LM. A media planning calculus. Operations Research. 1969;**17**(1):1-35. DOI: 10.1287/opre.17.1.1

[2] Seidmann TI, Sethi SP, Derzko N. Dynamics and optimization of a distributed sales-advertising model. Journal of Optimization Theory and Applications. 1987;**52**(3):443-462. DOI: 10.1007/BF00938216

[3] Grosset L, Viscolani B. Advertising for the introduction of an age-sensitive product. Optimal Control Applications & Methods. 2005;**26**(3):157-167. DOI: 10.1002/oca.758

[4] Buratto A, Grosset L, Viscolani B. Advertising a new product in segmented market. EJOR. 2006;**175**(2):1262-1267. DOI: 10.1016/j.ejor.2005.06.035

[5] Nerlove M, Arrow KJ. Optimal advertising policy under dynamic conditions. Economica. 1962;**29**(114): 129-142 http://www.jstor.org/stable/2551549

[6] Thompson GL, Teng JT. Optimal pricing and advertising policies for new product oligopoly market. Marketing Science. 1984;**3**(2):148-168. DOI: 10.1287/mksc.3.2.148

[7] Bass FM. A new product growth model for consumer durables. Management Science. 1969;**15**(5): 215-227. DOI: 10.1287/mnsc.15.5.215

[8] Robinson B, Lakhani C. Dynamic price models for new product planning. Management Science. 1975;**21**(10): 1113-1122. DOI: 10.1287/mnsc.21.10.1113

[9] Horsky D, Simmon L. Advertising and diffusion of new products. Marketing Science. 1983;**2**(1):1-17. DOI: 10.1287/mksc.2.1.1

[10] Kalish S. Monopolist pricing with dynamic demand and product cost. Marketing Science. 1983;**2**(2):135-159. DOI: 10.1287/mksc.2.2.135

[11] Kalish S. A new product adoption model with price, advertising and uncertainty. Management Science. 1985; **31**(12):1569-1585. DOI: 10.1287/mnsc.31.12.1569

[12] Sethi SP, Bass FM. Optimal pricing in a hazard rate model of demand. Optimal Control Applications & Methods. 2003;**24**(4):183-196. DOI: 10.1002/oca.729

[13] Kamakura W, Balasubramanium SK. Long-term view of the diffusion of durables: a study of the role of price and adoption influence processes via tests of nested models. International Journal of Research in Marketing. 1988;**5**(1):1-13. DOI: 10.1016/0167-8116(88)90012-2

[14] Dockner E, Jørgensen S. Optimal advertising policies for diffusion models of new product innovation in monopolistic situations. Management Science. 1988;**34**(1):119-130. DOI: 10.1287/mnsc.34.1.119

[15] Simon H, Sebastian K. Diffusion and advertising: German telephone company. Management Science. 1987; **33**(4):451-466. DOI: 10.1287/mnsc.33.4.451

[16] Thompson GL, Teng JT. Optimal strategies for general price-quality decision models of new products with learning production costs. EJOR. 1996; **93**(3):476-489. DOI: 10.1016/0377-2217(96)00028-8

[17] Bass FM, Krishnan TV, Jain DC. Why the Bass model fits without decision variables. Marketing Science. 1994;**13**(3):203-223. DOI: 10.1287/mksc.13.3.203

[18] Jha PC, Chaudhary K, Kapur PK. Optimal advertising control policy for a new product in segmented market. Opsearch. 2009;**46**(2):225-237. DOI: 10.1007/s12597-009-0014-1

[19] Manik P, Chaudhary K, Singh Y, Jha PC. Optimal promotion effort control policy for segment specific new product growth. In: Bansal JC, Singh PK, Deep K, Pant M, Nagar A, editors. Advances in Intelligent and Soft Computing, Vol 202. India: Springer; 2013. pp. 347-358. DOI: 10.1007/978-81-322-1041-2_30

[20] Huang J, Leng M, Liang L. Recent developments in dynamic advertising research. EJOR. 2012;**220**(3):591-609. DOI: 10.1016/j.ejor.2012.02.031

[21] Helmes K, Schlosser R, Weber M. Optimal advertising and pricing in a class of general new-product adoption models. EJOR. 2013;**229**(2):433-443. DOI: 10.1016/j.ejor.2013.02.035

[22] Seierstad A, Sydsaeter K. Optimal Control Theory with Economic Applications. North-Holland, Amesterdam; 1987

[23] Sethi SP, Thompson GL. Optimal Control Theory: Applications to Management Science and Economics. Dordrecht: Kluwer Academic Publishers; 2000

[24] Teng JT, Thompson GL. Oligopoly models for optimal advertising when production costs obey a learning curve. Marketing Science. 1983;**29**(9): 1087-1101. DOI: 10.1287/mnsc.29.9.1087

[25] Rosen JB. Numerical solution of optimal control problems. In: Dantzig GB, Veinott AF, editors. Mathematics of Decision Science, Part-2. American Mathematical Society; 1968. pp. 37-45

[26] Thirez H. OR software LINGO. EJOR. 2000;**124**:655-656. DOI: 10.1016/0377-2217(95)90045-4

Chapter 3

Integrated Batch Production and Maintenance Scheduling to Minimize Total Production and Maintenance Costs with a Common Due Date Constraint

Zahedi Zahedi

Abstract

This chapter discusses an integrated model of batch production and machine maintenance scheduling on a single deteriorating machine and flow shop with a deteriorating machine that produces an item to be delivered at a common due date. The model describes the trade-off between production costs and maintenance costs as the increase of production run length. The objective function of the model is to minimize total cost consisting of in-process and complete inventory holding costs, setup cost, preventive and corrective maintenance costs, and rework cost. The problem is to determine the best production run length and maintenance actions that minimize the total cost.

Keywords: batch production, maintenance scheduling, single item, deteriorated machine

1. Introduction

The development of today's manufacturing systems leads to a shorter product life cycle, increasing product varieties and customer demand on the higher quality and timeliness of delivery. Thus, the accuracy and speed of decision-making in the manufacturing system become important.

A machining process industry gets many machining orders from its strategic partner industries in large quantities. The manufacturer processes the order in constant batch size that is set by production section. Meanwhile, Maintenance Section performs machine maintenance only in case of machine failures (reactive maintenance). Delay delivery order to consumers cannot be avoided if the maintenance machine takes a long time and disrupts production activities, and this often happens.

Some root of the problem can be drawn from this description: first, the maintenance has not implemented preventive maintenance system although the machine failure data, the time interval between failures, and the cost of each failure are well recorded. Second, the production section schedules batches in a constant size, whereas according to [1–6], discussions about nonconstant batch sizes will provide a better shop time. Third, the machine failure occurs when production is in progress, so the machine failure interferes with the productivity of the shop floor. These three issues indicate the independence between production scheduling, and maintenance scheduling may result in the following conditions:

1. Production scheduling that does not take into consideration the maintenance aspect will cause a machine to be continually operated even though the machine should be maintained. If the maintenance is not conducted, it can result in a breakdown machine during production activities that will certainly interfere with productivity.

2. Maintenance schedule that does not take into account the production schedule will cause a busy machine to be stopped for maintenance. This also disrupts the pre-arranged production schedule.

2. Model construction

This section discusses how the model is constructed. The linkage between the uncontrollable parameter and the model decision variable in achieving the best total cost is discussed in the influence diagram for integrated batch production scheduling and maintenance scheduling, input-output diagram for the models of batch production and maintenance scheduling, construction of objective function, system constraint, model and algorithm, and then a numerical experience to show how an algorithm works to solve a problem.

2.1 *Influence diagram* for the model batch production and maintenance scheduling

The integrated batch production and maintenance scheduling developed in this chapter have total cost minimization criteria consisting of inventory holding cost, setup cost, PM cost, CM cost, and rework cost for nonconforming part. In the influence diagram **Figure 1**, the problem of integrated batch production and maintenance scheduling can be explained as follows. Demand, due date, and machine performances are uncontrollable parameters. Demand and due date will affect batch size (production schedule).

The machine performance will affect the estimated number of PM and number of CM. In the model system, the number of PM and number of CM will affect each other with the production schedule. The number of PM will affect each other with the number of CM, where the increasing number of PM will cause the number of CM to decrease and vice versa. The number of PM and CM also influences each other for the number of nonconforming parts.

2.2 Input-output diagram for the models of batch production and maintenance scheduling

The total cost consists of inventory holding costs, setup cost, PM cost, CM cost and rework cost. The cost of inventory holding costs consists of the work in process (wip) and the finished part of inventory holding costs. In-process inventory holding cost is the cost of inventory for parts in batches during processing are calculated by multiplying the number of parts in the batch by the waiting time for the batch to be processed. The cost of the finished part is the inventory holding cost for parts in the finished batch, calculated by multiplying the number of parts in the batch by unit inventory holding cost for finished part as long as the time of waiting for the batch

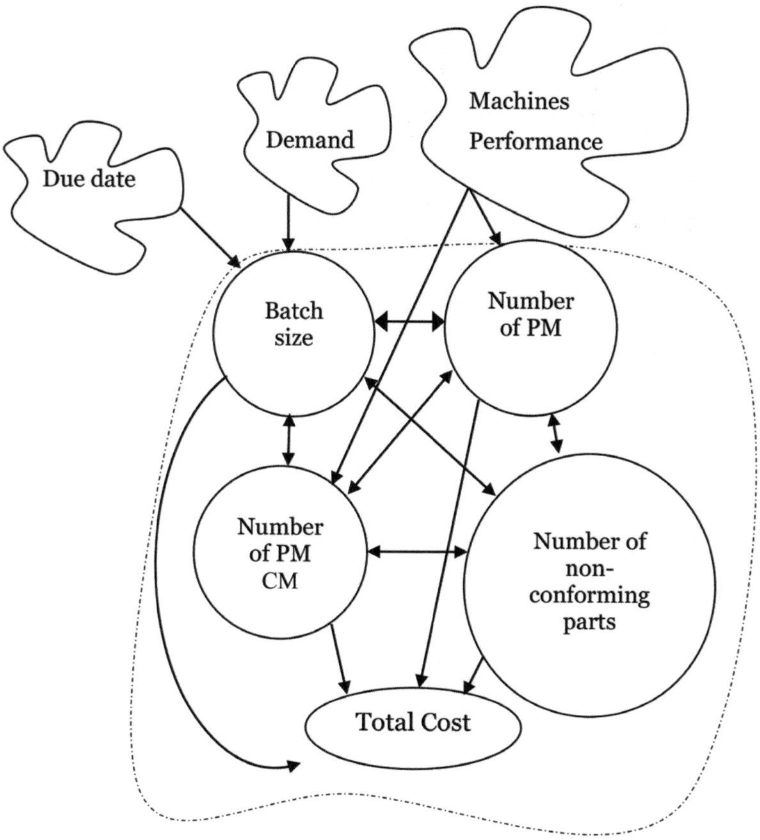

Figure 1.
Influence diagram for the model of batch production and maintenance scheduling.

until the deadline d. The setup cost is calculated by multiplying the number of scheduled batches by unit setup cost. The rework cost is calculated by multiplying the number of nonconforming parts by unit rework cost. The PM cost is calculated by multiplying the number of PM by unit PM cost, and the CM cost is calculated by multiplying the number of CM by unit CM cost.

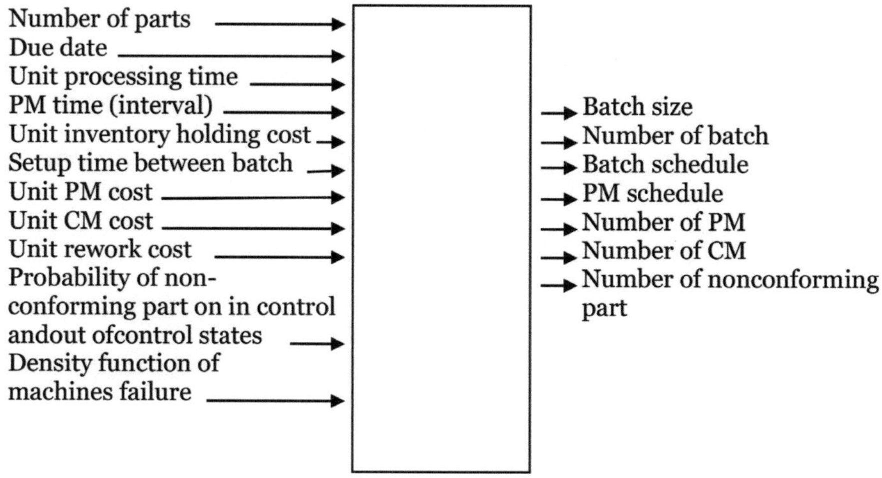

Figure 2.
Input-output diagram for the models of batch production and maintenance scheduling.

The input-output diagram of the integrated batch production and maintenance scheduling to minimize total cost is shown in **Figure 2**.

Figure 2 shows the input parameters model: the number of parts scheduled, due date, Weibull distribution function $f(t)$, unit processing time, PM interval length, inventory holding cost, rework cost, setup time between batches, and the probability of nonconforming part on the machine in the status of in-control and out of control. The output of the model are the size of batches and the schedule, PM schedule, number of CM, and number of nonconforming parts.

The model will address trade-off issues on production costs and maintenance costs, where production costs will consist of inventory holding costs (in-process and finished part of inventory holding costs), setup cost, and rework cost for nonconforming parts, while the maintenance cost consists of PM cost and CM cost. The model will answer how the batch production and maintenance scheduled minimize the total cost. Drawing influence diagram follows [7].

3. Integrated batch production and maintenance scheduling for single item processed on a deteriorating machine with a due date

The modeling starts with an inventory holding cost for in-process and completed batches, modeling of system constraints of the problem, model of the problem, algorithm, and an example to show how the algorithm works to solve the model.

3.1 Inventory holding costs for in-process batch and completed batch

Holding cost concept is developed from [3]. Let q parts of an item be scheduled by a minimization of total actual flow time criterion. The q parts are divided into N batches $L_{[i]}$ $(i = 1, 2, \ldots, N)$ where the sizes of each batch are $Q_{[i]}$ $(i = 1, 2, \ldots, N)$. If all parts in a batch have been processed completely, then the batch is called as completed batch. If a batch is still containing any part not yet or being processed, it is called as in-process batch.

In an assumption raw materials arrive just in the time when they are required, i.e., in the beginning of a batch processing, the holding cost is only for in-processed batch and completed batch. The formulation of holding cost is conducted for in-process batch firstly and then for completed batch. The position of *batch* $L_{[i]}$ in a single machine manufacturing system by a backward approach during a planning horizon is shown in **Figure 3**.

An assumption for in-process batch is that parts in a batch shall wait in the batch until all parts in the batch have already been processed. Therefore, in the interval

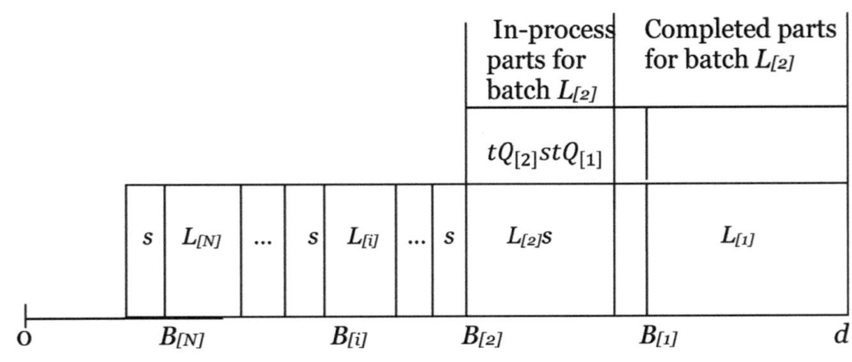

Figure 3.
Batch position in a single machine manufacturing system.

$(0,t)$ in the in-process batch $L_{[i]}$, there are $Q_{[i]}$ work-in-process parts (parts not yet or being processed). In interval $(t, 2\,t)$ there are $(Q_{[i]}\text{-}1)$ work-in-process parts and l finished part, until in interval $((Q_{[i]}-1)t, Q_{[i]}t)$, there are I work-in-process parts and $(Q_{[i]}-1)$ finished parts.

The amount of holding cost for the finished part, designated f_1, is as follows:

$$f_1 = c_1t + c_12t + c_13t + \ldots + c_1t\left(Q_{[i]}-3\right) + c_1t\left(Q_{[i]}-2\right) + c_1t\left(Q_{[i]}-1\right).$$

The amount of holding cost for the in-process part, designated f_1, is as follows:

$$f_2 = c_2tQ_{[i]} + c_2t\left(Q_{[i]}-1\right) + c_2t\left(Q_{[i]}-2\right) + \ldots + c_23t + c_22t + c_2t.$$

By summation f_1 and f_1, f_2 and f_2 in a reverse order, a simpler result is found, i.e.:

$$\mathrm{f}_1 = \frac{c_1}{2}\left[tQ_{[i]}\left(Q_{[i]} - 1\right]\mathrm{danf}_2 = \frac{c_2}{2}\left[t\left(Q_{[i]} + 1\right)\left(Q_{[i]}\right]\right..$$

Then, the holding cost of ith in-process batch is the addition of f_1 and f_2, that is:

$$f_1 + \mathrm{f}_2 = \frac{c_1}{2}\left[tQ_{[i]}\left(Q_{[i]} - 1\right] + \frac{c_2}{2}\left[t\left(Q_{[i]} + 1\right)\left(Q_{[i]}\right]\right.\right. \text{ or}$$

$$f_1 + f_2 = \frac{c_1 + c_2}{2}tQ_{[i]}^2 + \frac{c_2 - c_1}{2}tQ_{[i]} \tag{1}$$

Based on Eq. (1), the total holding cost of the in-process batch of all batches may be written as follows:

$$\frac{c_1 + c_2}{2}tQ_{[N]}^2 + \frac{c_2 - c_1}{2}tQ_{[N]} + \frac{c_1 + c_2}{2}tQ_{[N-1]}^2 + \frac{c_2 - c_1}{2}tQ_{[N-1]} + \ldots$$
$$+\frac{c_1 + c_2}{2}tQ_{[2]}^2 + \frac{c_2 - c_1}{2}tQ_{[2]} + \frac{c_1 + c_2}{2}tQ_{[1]}^2 + \frac{c_2 - c_1}{2}tQ_{[1]} = \frac{c_1 + c_2}{2}t\sum_{i=1}^{N}Q_{[i]}^2 + \frac{c_2 - c_1}{2}t\sum_{i=1}^{N}Q_{[i]} \tag{2}$$

The holding cost of completed batch may be formulated as follows:

$$\mathrm{c}_1\left\{\left(tQ_{[N-1]} + s\right) + \ldots + \left(tQ_{[2]} + s\right) + \left(tQ_{[1]} + s\right)\right\}Q_{[N]}$$
$$+\mathrm{c}_1\left\{\left(tQ_{[N-2]} + s\right) + \ldots + \left(tQ_{[2]} + s\right) + \left(tQ_{[1]} + s\right)\right\}Q_{[N-1]} + \ldots$$
$$+\mathrm{c}_1\left\{\left(tQ_{[2]} + s\right) + \left(tQ_{[1]} + s\right)\right\}Q_{[3]} + \mathrm{c}_1\left\{\left(tQ_{[1]} + s\right)\right\}Q_{[2]} = \mathrm{c}_1\sum_{i=1}^{N-1}\left\{\sum_{j=1}^{i}\left(tQ_{[j]} + s\right)\right\}Q_{[i+1]}. \tag{3}$$

Next, total holding cost ($TolC$) is computed by adding up the holding cost of in-process batch in Eq. (2) and completed batch in Eq. (3) to yield Eq. (4).

$$TolC = \mathrm{c}_1\sum_{i=1}^{N-1}\left\{\sum_{j=1}^{i}\left(tQ_{[j]} + s\right)\right\}Q_{[i+1]} + \frac{c_1 + c_2}{2}t\sum_{i=1}^{N}Q_{[i]}^2 + \frac{c_2 - c_1}{2}t\sum_{i=1}^{N}Q_{[i]}. \tag{4}$$

The first term of Eq. (4) is total holding cost in completed batch, and the second and third terms are the total holding cost, while the part is being processed in batch (in-process batch) in one production run.

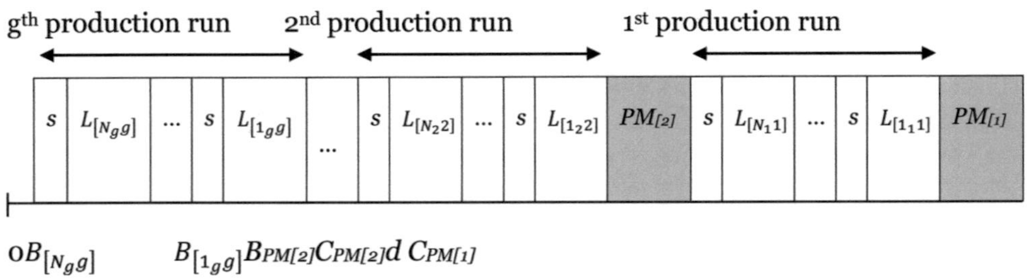

Figure 4.
Batch position in a single machine manufacturing system.

Eq. (4) and **Figure 3** are to be developed into a formulation of holding cost for g production run and PM interval inserted sequentially as shown in **Figure 4**.

By considering any changes taking place in each production run and total PM for g production runs and g PM intervals, total holding cost will become Eq. (5).

$$c_1 \sum_{i=1}^{N-1} \left\{ \sum_{j=1}^{i} \left(tQ_{[j_1 1]} + s \right) \right\} Q_{[(i+1)_1 1]} + \frac{c_1 + c_2}{2} t \sum_{i_1=1}^{N_1} Q_{[i_1 1]}^2 + \frac{c_2 - c_1}{2} t \sum_{i_1=1}^{N_1} Q_{[i_1 1]} +$$
$$\sum_{k=2}^{g} \left[c_1 \sum_{i_k=1}^{N_k - 1} \left\{ \sum_{j=1}^{i} \left(tQ_{[j_k k]} + s \right) \right\} Q_{[(i+1)_k k]} + \frac{c_1 + c_2}{2} t \sum_{i_k=1}^{N_k} Q_{[i_k k]}^2 + \right.$$
$$\left. \frac{c_2 - c_1}{2} t \sum_{i_k=1}^{N_k} Q_{[i_k k]} + c_1 \sum_{i_k=1}^{N_k} Q_{[i_k k]} \left((k-1) t_{PM} + \sum_{j_k=1}^{N_{(k-1)}} \left(tQ_{[j_k k]} + s \right) \right) \right]. \tag{5}$$

3.2 ROCOF function

Rate of occurrence of failures (ROCOF) is a concept that is useful in the modeling of failures over time and the effect of PM (and CM) actions [8]. The ROCOF characterizes the probability that a failure occurs in the interval $[t, t + \delta t]$. The ROCOF is given by an intensity function

$$\lambda(t) = \lim_{\delta t \to 0} \frac{P\{N(t + \delta t) - N(t) \geq 1\}}{\delta t} \tag{6}$$

where N(t) is the number of failures in the interval [0,t). Since the probability of two or more failures in the interval $[t, t + \delta t]$ is zero as $\delta t \to 0$, we have the intensity function equal to the derivative of the conditional expected number of failures, so that

$$\lambda(t) = \frac{d}{dt} E\{N(t)\}. \tag{7}$$

When the failures are minimally repaired and the time to repair is negligible, then ROCOF function $\lambda(t) = r(t)$, the failure rate function. The cumulative ROCOF function is given by

$$\Lambda(t) = \int_0^t \lambda(t) dt \tag{8}$$

A ROCOF function that has been used extensively is the Weibull ROCOF. The cumulative ROCOF (or the expected total number of failures) is given by the function

$$\Lambda(t) = \left(\frac{t}{\alpha}\right)^{\beta} \tag{9}$$

with scale parameter α and shape parameter β.

Let a system (machine) with a Weibull failure time distribution have a shape parameter of β = 1.69 and a scale parameter α = 2,857.14, then, based on ROCOF cumulative function, the first, the second, and so on estimated failure times that could be written as follows can be found:

If $\Lambda(t) = \left(\frac{t}{\alpha}\right)^{\beta} = 1$ then t = 2857.14. If $\Lambda(t) = \left(\frac{t}{\alpha}\right)^{\beta} = 2$ then t = 4305.82. If $\Lambda(t) = \left(\frac{t}{\alpha}\right)^{\beta} = 3$ then t = 5473.33. If $\Lambda(t) = \left(\frac{t}{\alpha}\right)^{\beta} = 4$ then t = 6489.03.

From the calculation above, the time interval between machine failure times can be estimated, where the time between failures of a machine is diminishing over time. It indicates that the machine has increasing failure rate distribution.

3.3 Estimation of nonconforming parts

This research developed a policy in that PM is carried out before an expected first failure time based on cumulative ROCOF function. An example of a condition for a case of two production runs and two PMs is shown in **Figure 5**. In the second production run, there is no nonconforming part, because the out-of-control state takes place in the first production run, so that the number of nonconforming parts for k = 2 may be written as follows:

$$M_2 = p_2 \text{ xnumber of parts processed in interval} \left[B_{[N_1 1]} - s + \alpha, C_{[11]}\right] \tag{10}$$

In the same way for g production runs and g PMs, the number of nonconforming parts will always be of the same form, except if applied to k = 1, 2, ..., g, so that

$$M_g = p_2 \text{ xnumber of parts processed in interval} \left[B_{[N_1 1]} - s + \alpha, C_{[11]}\right] \tag{11}$$

Under an assumption that the probability that the nonconforming part processed under in-control state is p_1 = 0, then the expected number of nonconforming parts may be written as

$$E(M) = M_k, k = 1, 2, \ldots, g \tag{12}$$

so that the expected network cost may be computed by

$$E(W) = c_w E(M) \tag{13}$$

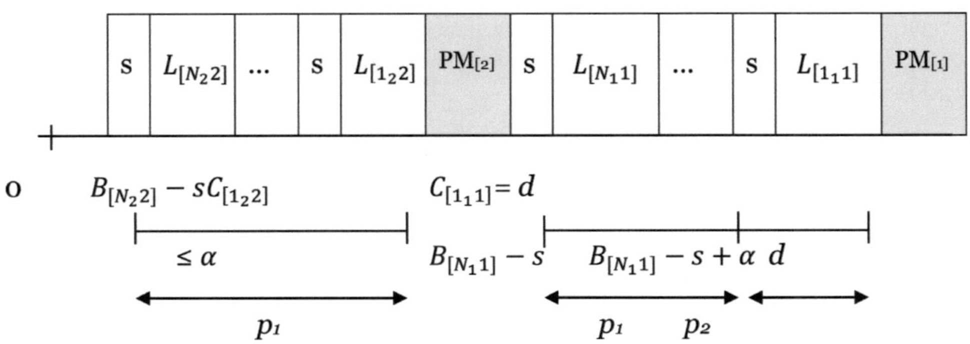

Figure 5.
A condition for two production runs and two PMs.

3.4 Model formulation

In order to formulate the integrating batch production and maintenance scheduling into a mathematical model, we use following notations.

Parameters

t: unit processing time for a part
s: setup time required for any batch processed
c_1: unit inventory holding cost for finished part per unit per time unit
c_2: unit inventory holding cost for the work in process part per unit per time unit
c_s: unit setup cost
c_{PM}: unit preventive maintenance cost
c_r: unit restoration cost (corrective cost)
c_w: unit rework cost per part for nonconforming part
t_{PM}: time interval for preventive maintenance (in constant assumption)
β: shape parameter for the Weibull distribution
α: scale parameter for the Weibull distribution
p_1: probability of defect part on in-control state
p_2: probability of defect part on out-of-control state
q: number of parts to be processed
d: an order delivery time (a common due date)

Decision variables

g: number of production run in model [SISM]
R: number of production run in algorithm [SISM]
$L_{[i_k k]}$: batch scheduled at position i in the kth production run from due date direction (backward approach), for $i_k = 1, 2, \ldots, N_k$, $k = 1, 2, \ldots, g$
$Q_{[i_k k]}$: batch size of $L_{[i_k k]}$
N: possible total number of batch in a planning horizon
N_k: number of batch in k^{th} production run, $k = 1, 2, \ldots, g$
$B_{[i_k k]}$: beginning time for batch $L_{[ik]}$
$C_{[i_k k]}$: completion time for batch $L_{[ik]}$
$B_{PM[k]}$: beginning time for k^{th} PM
$C_{PM[k]}$: completion time for k^{th} PM

$$X_{[i_k k]} = \begin{cases} 1, \text{if} Q_{[i_k k]} \neq 0, \\ 0, \text{if} Q_{[i_k k]} = 0 \end{cases}, \; i = 1, 2, \ldots, N_k,\; k = 1, 2, \ldots, g$$

n_{CM}: number of CM minimal repair (restoration)
R: total cost of CM (restoration)
M: number of nonconforming parts
Objective function TC: the total cost consisting of inventory cost in process and complete inventory costs, setup cost, preventive and corrective maintenance cost, and rework cost

The model has some assumptions in formulating the model, as follows:

1. This integrating model for single item processed on a single deteriorating machine.

2. Setup time is not depending on the size of batches.

3. Batch position number and PM are counted from due date direction (backward approach).

4. The same load force for machine in setup time and in processing time.

5. The machine cannot be interrupted as long as production runs.

6. Batch size value is in real positive.

Using those defined notations and based on those assumptions, the integrating batch production and maintenance scheduling to minimize production and maintenance costs on a deteriorating machine in just in time environment (model [SISM]) can be expressed as mixed-integer nonlinear programming as follows:

Model [SISM]

$$\text{Minimize } TC = c_1 \sum_{i_1=1}^{N_1-1} \left\{ \sum_{j_1=1}^{i_1} \left(tQ_{[j_1 1]} + s \right) \right\} Q_{[(i+1)_1 1]} + \frac{c_1 + c_2}{2} t \sum_{i_1=1}^{N_1} Q^2_{[i_1 1]} +$$

$$\frac{c_2 - c_1}{2} t \sum_{i_1=1}^{N_1} Q_{[i_1 1]} + \sum_{k=2}^{g} \left[c_1 \sum_{i_k=1}^{N_k-1} \left\{ \sum_{j_k=1}^{i_k} \left(tQ_{[j_k k]} + s \right) \right\} Q_{[(i+1)_k k]} + \frac{c_1 + c_2}{2} t \sum_{i_k=1}^{N_k} Q^2_{[i_k k]} + \right.$$

$$\left. \frac{c_2 - c_1}{2} t \sum_{i_k=1}^{N_k} Q_{[i_k k]} + c_1 \sum_{i_k=1}^{N_k} Q_{[i_k k]} \left((k-1)t_{PM} + \sum_{j_k=1}^{N_{(k-1)}} \left(tQ_{[j_k k]} + s \right) \right) \right] + gc_{PM} +$$

$$c_s \sum_{k=1}^{g} N_k + \mathrm{E(R)} + E(W) \tag{14}$$

subject to

$$\sum_{k=1}^{g} \sum_{i_k=1}^{N_k} Q_{[i_k k]} = q \tag{15}$$

$$B_{[i_1 1]} + \sum_{j_1=1}^{i_1} \left(sX_{[j_1 1]} + tQ_{[i1]} \right) - s = d, i_1 = 1, \ldots, N_1, k = 1 \tag{16}$$

$$B_{[i_k k]} + \sum_{l=2}^{k} \left[\sum_{j_l=1}^{i_l} \left(sX_{[j_l l]} + tQ_{[j_l l]} + (k-1)t_{PM} \right) \right] - s +$$

$$\sum_{i_1=1}^{N_1} \left(sX_{[i_1 1]} + tQ_{[i_1 1]} \right) = d, i_k = 1, 2, \ldots, N_k \text{ and } k = 2, 3, \ldots, g \tag{17}$$

$$\sum_{i_k=1}^{N_k} \left(tQ_{[i_k k]} + s \right) \le d, k = 1 \tag{18}$$

$$\sum_{i_k=1}^{N_k} \left(tQ_{[i_k k]} + s \right) \le \alpha, k = 2, 3, \ldots, g \tag{19}$$

$$M_k = p_2 \text{xnumber of parts processed in interval}$$

$$\left[B_{[N_1 1]} - s + \alpha, C_{[11]} \right], k = 1, 2, \ldots, g \tag{20}$$

$$E(M) = M_k, k = 1, 2, \ldots, g \tag{21}$$

$$\mathrm{E(W)} = \mathrm{c_w} E(M) \tag{22}$$

$$n_{CM} = \left\lfloor \left\{ \frac{d - \left(B_{[N_1 1]} - s \right)}{\alpha} \right\}^{\beta} \right\rfloor \tag{23}$$

$$E(R) = c_r n_{CM} \tag{24}$$

$$B_{PM\ [1]} = d,$$

$$C_{PM[1]} = d + t_{PM},$$

$$B_{PM[k]} = B_{[1_k k]} + tQ_{[1_k k]}, k = 1, 2, \ldots, g,$$

$$C_{PM[k]} = B_{PM[k]} + t_{PM}, k = 1, 2, \ldots, g \quad (25)$$

$$N = \left\lfloor \frac{d - \left\lceil \frac{d}{\alpha} \right\rceil t_{PM} - tq}{s} \right\rfloor \quad (26)$$

$$g = \lceil tq/\alpha \rceil \quad (27)$$

$$X_{[i_k k]} = \begin{cases} 1, \text{if} Q_{[i_k k]} \neq 0, \\ 0, \text{if} Q_{[i_k k]} = 0 \end{cases}, i_k = 1, 2, \ldots, N_k, k = 1, 2, \ldots, g \quad (28)$$

$$Q_{[i_k k]} \geq 0, i_k = 1, 2, \ldots, N_k, k = 1, 2, \ldots, g \quad (29)$$

$$Q_{[i_k k]} \leq X_{[i_k k]} q, i_k = 1, 2, \ldots, N_k, k = 1, 2, \ldots, g \quad (30)$$

$$N_k \geq 1, \mathrm{k} = 1, 2, \ldots, g \quad (31)$$

Eq. (14) declares an objective function to minimize total costs consisting of inventory costs, setup cost, preventive maintenance cost, corrective maintenance cost, and rework cost. Eq. (15) states the balance of the material in the shop, where the number of parts in all batches must be equal to the number of parts that will be scheduled. Eqs. (16) and (17) state the beginning time of each batch on the first run and the next runs, respectively. All batches are scheduled tight to a common due date d sequentially. Eqs. (18) and (19) state the length of first run and the next runs, respectively. Eq. (20) states the estimation of nonconforming parts for each run. Eqs. (21) and (22) state the estimation of total nonconforming parts and total rework cost for nonconforming parts, respectively. Eqs. (23) and (24) state the possible number of CM action with cumulative Weibull ROCOF and the expected cost of CM action, respectively. Eq. (25) represents a set of constraints for the beginning and the next of the PM times, with the assumption that first PM in schedule or the last PM in processing (backward approach) after all batches has been completed at a common due date d to ensure the machine in as good as new condition for the next order. Eq. (26) states the possible number of batches in a planning horizon. Eq. (27) states the possible number of production runs in a planning horizon. Equation (28) states a binary constraint that each batch will have: $X_{[i_k k]} = 1$ for non-empty batches and $X_{[i_k k]} = 0$ for empty batches. Eq. (29) states non-negativity of batch size. Eq. (30) states batch size less or equal with all parts that will be scheduled. Eq. (31) states the existence of the number of batches in each run.

3.5 Algorithm

The algorithm developed begins with problem solving without involving restoration cost (CM) and rework cost for nonconforming parts or without the constraints of Eqs. (20)–(24). It begins with one batch in one production run with one PM. After having obtained a production schedule, estimate the number of nonconforming parts by Eq. (20) and the number of restoration (CM) by Eq. (23). Next, compute estimated rework cost by Eq. (22) and estimated restoration cost by Eq. (24), and then compute total cost. This step is done for two batches until an increased total cost is found. Write the best total cost for one production run and one PM. This process is carried out for two production runs with two PMs until the best total cost is found for two production runs with two PMs. Continue the process up to g production runs with g PMs. The best algorithm solution is the minimization of all the best total costs for k = 1, 2, ..., g. Then, write all decision variables of single item single machine integration problem. Single item single machine integration problem algorithm fully is as SISM algorithm.

Algorithm [SISM]

Step-1. Set the length of the first failure time interval after PM as α. Go to Step -2.
Step-2. A problem is said as feasible if and only if the total time of process with one setup doesn't exceed the due date of delivering d, otherwise the problem is unfeasible for a model or if $s + tq \le d$ then the problem is feasible; continue to Step-3. If $s + tq > d$, the problem is unfeasible, stop.
Step-3. Compute g by Eq. (27), and set $N_k = \lfloor N \rfloor$, being computed by Eq. (26), $k = 1, 2, \ldots, g$. Go to Step-4.
Step-4. For R = 1, 2, ..., g. Go to Step-5.
Step-5. Set R = *1*. Go to Step-6.
Step-6. Set g = R. Go to Step-7.
Step-7. Substitute the values of g, N_k, p, q, t, s, d, t_{PM} into the model and set Set $X_{[i_k k]} = 1$ for $i_k=1$ and $k=1$ dan set $X_{[i_k k]} = 0$ for other i_k and k. Go to Step-8.
Step-8. Solve SISM Model without the constraint of Eqs. (20)-(24). Compute estimated rework cost by Eq. (22) and estimated restoration cost by Eq. (24), and compute a total cost to find TC, write $TC_{[1,1]} = TC$. Go to Step-9.
Step-9. Set $k = 1$. Go to Step-10.
Step-10. Set $i_k = 2$. Go to Step-11.
Step-11. Set $X_{[j_l k]} = 1$ for $j_l = 1, 2, \ldots, i_k$ and $l = 1, 2, \ldots, k$, and $X_{[j_l k]} = 0$ otherwise. Go to Step-12.
Step-12. Solve SISM Model without the constraint of Eqs. (20)-(24). Compute estimated rework cost by Eq. (22) and estimated restoration cost by Eq. (24), and compute a total cost to find TC, write $TC_{[i_k k]} = TC$. Go to Step-13.
Step-13. Observe whether $TC_{[i_k k]} < TC_{[(i-1)_k k]}$,
- If $TC_{[i_k k]} < TC_{[(i-1)_k k]}$, observe whether $i_k = N_k$,
  - If $i_k = N_k$, go to Step -14.
  - If $i_k \neq N_k$, set $i_k = i_k + 1$, go back to Step-11.
- If $TC_{[i_k k]}$ $TC_{[(i-1)_k k]}$, write $TC_{[k]}{}^* = TC_{[(i-1)_k k]}$ and writeall of $TC_{[k]}{}^*$ *-related decision variables*, go to Step-15.

Step-14. Write $TC_{[k]}{}^* = TC_{[i_k k]}$ and writeall of TC^**-related decisionvariables*, go to Step-15.
Step-15. Observe whether $k = g$,
  - if $k = g$, go to Step-21.
  - if $k \neq g$, go to Step-16.

Step-16. Set k = k + 1, go to Step-17.
Step-17. Set $i_k = 2$, go to Step-18.
Step-18. Set $X_{[j_l l]} = 1$ for $j_l = 1, 2, \ldots, i_k$ and $l = 1, 2, \ldots, k+1$, and set $X_{[j_l l]} = 0$ for other j_l, l. Go to Step-19.
Step-19. Solve SISM Model without the constraint of Eqs. (20)-(24). Compute estimated rework cost by Eq. (22) and estimated restoration cost by Eq. (24), and compute a total cost to find TC, write $TC_{[i(k+1)]} = TC$. Go to Step-20.
Step-20. Observe whether $TC_{[i_{(k+1)}(k+1)]} < TC_{[k]}{}^*$,
- If $TC_{[i_{(k+1)}(k+1)]} < TC_{[k]}{}^*$, Set $i = i + 1$, go back to Step-11.
- If $TC_{[i_{(k+1)}(k+1)]}$ $TC_{[k]}{}^*$, observe whether $k = g$,
  - If $k = g$, go to Step-21.

- If $k \neq g$,, Set $k = k + 1$, go back to Step-10.

Step-21. Write $TC_{[R]} = TC_{[k]}^*$, R = 1, 2, ..., g. Go to Step 22.

Step 22. Observe whether R = g

- If R = g. Go to Step 23.
- If R $\neq g$., Set R = R +1, go back to Step-6.

Step 23. Write $\{TC_{[R]}$, R = 1, 2, ..., g}.

Step-24. The minimal solution is

Min $\{TC_{[R]}$, R = 1, 2, ..., g}. Go to Step-25.

Step-25. Write all values of decision variables.

3.6 Numerical experience

To clarify how the proposed algorithm works, the following example is given. Consider an integrating problem with parameters as follows: number of parts q = 300 units, the setup time between batches s = 30 min, unit processing time of part t = 20 min, the length of the preventive maintenance time t_{PM} = 60 min = $1/\mu$ (constant), the shape parameter of Weibull distribution β = 1.69, and scale parameter α = 2857.14, constant repair rate μ = 1/60, common due date d = 10000.00, the unit inventory holding cost of finished parts c_1 = US$ 0.20 per unit per minute, the unit inventory holding cost of in-process parts c_2 = US$ 0.10 per unit per minute, the unit cost of PM c_{PM} = US$ 30.00, unit setup cost c_s = US$ 3.00, probability of defect part on in-control state p_1 = 0.00, probability of defect part on out-of-control state p_2 = 0.30, unit rework cost per unit part for nonconforming parts c_w = US $100.00, and unit corrective maintenance cost c_r = US$ 120.00.

The computational steps to solve the problem are the followings.

Step-1. Yields α = 2,857.14.

Step-2. Yields 30 + 20 x 300 = 6,030.00 ≤10,000.00 is met, then the problem is feasible.

Step-3.Yields g = 3 and N_k = 125, k = 1, 2, 3.

Step-4. Yields R = 1, 2, 3.

Step-5 to Step-21, for R = 1, yield the best solution $TC_{[1]}^*$ = 201,313.00.

1st looping

Step-5 to Step-21, for R = 2, yield the best solution $TC_{[2]}^*$ = 201,124.80.

2nd looping

Step-5 to Step-21, for R = 3, yield the best solution $TC_{[3]}^*$ = 201,158.80.

Step-23. Yields a set of the best solutions for R = 1, 2, 3 as a set of $\{TC_{[1]}^*, TC_{[2]}^*, TC_{[3]}^*\}$ = {201,313.00, 201,124.80, 201,158.80}.

Step-24. Yields a minimal TC = min {201,313.00, 201,124.80, 201,158.80} = 201,124.80 that occurs in two production runs with two PMs.

Step-25. Yield the complete solution as shown in **Table 1** and **Figure 6**.

3.7 Conclusion

This chapter proposes a model of integrating batch scheduling and maintenance scheduling by criterion of minimization in holding cost, setup cost, PM cost, rework cost, and restoration cost (CM). The criterion of scheduling used is the minimization of the total actual flow time. The first preventive maintenance (PM) scheduling policy (from the direction of due date) is made precisely on due date. The second PM and so on would be done before the time of first deterioration according to Weibull ROCOF cumulative function.

In the model developed, searching for a solution begins with problem resolving without involving restoration cost (CM) and rework cost for nonconforming part. It begins with one batch in one production run and one PM. After finding a

$L_{[i_k k]}$	$Q_{[i_k k]}$	$B_{[i_k k]}$	$B_{PM[k]}$	$C_{PM[k]}$	$TC_{[2]}^*$
$L_{[1_1 1]}$	42.46	9150.77	$B_{PM[1]}$ = 10000.00	$C_{PM[1]}$ = 10060.00	201124.80
$L_{[2_1 1]}$	39.46	8331.54			
$L_{[3_1 1]}$	36.46	7572.31			
$L_{[4_1 1]}$	33.46	6873.08			
$L_{[5_1 1]}$	30.46	6233.85			
$L_{[6_1 1]}$	27.46	5654.62			
$L_{[7_1 1]}$	24.46	5135.39			
$L_{[8_1 1]}$	21.46	4676.15			
$L_{[9_1 1]}$	18.46	4276.92			
$L_{[10_1 1]}$	15.46	3937.69	$B_{PM[2]}$ = 3847.69	$C_{PM[2]}$ = 3907.69	
$L_{[1_2 2]}$	6.46	3718.46			
$L_{[2_2 2]}$	3.46	3619.23			
$L_{[3_2 2]}$	0.46	3580.00			

Table 1.
Best schedule for the example.

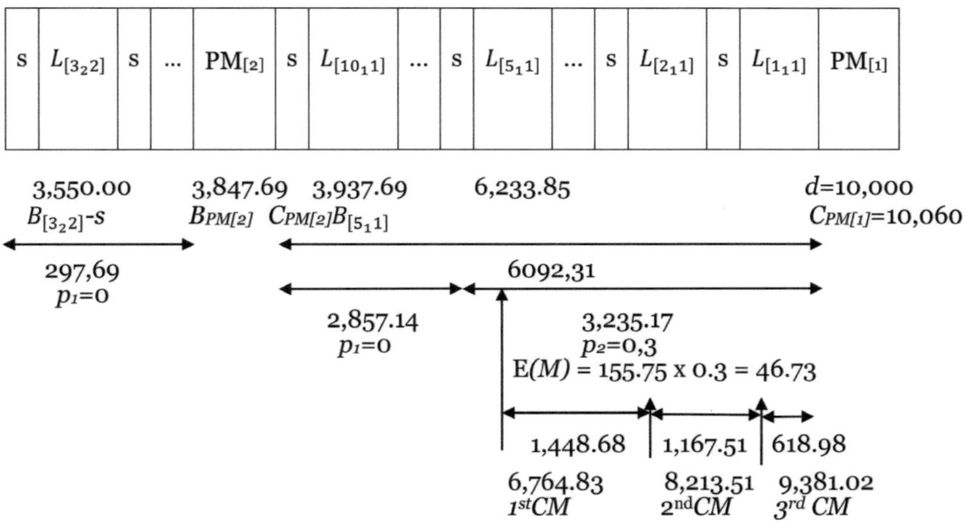

Figure 6.
Gantt chart of the best solution for the example.

production schedule, estimate the number of nonconforming parts and total restoration (CM). Next, compute estimated rework cost and estimated restoration cost, and then compute total cost. This step is done for two batches and so on until an increase on total cost is found. Write the best total cost for one production run and one PM. This process is carried out for two production runs with two PMs until the best total cost is found for two production runs with two PMs. Continue the process until g production runs with g PMs. The best solution of algorithm is a minimization of all of the best total cost for $k = 1, 2, \ldots, g$.

The model makes a trade-off in the following two things. An increase in the number of batch (length of production run) up to a certain limit will minimize the total inventory holding cost. Meanwhile, an increase in the length of production run will imply on an increase in the number of nonconforming parts and in a number of restoration (CM).

For readers who want to learn more about the integration of batch production and machine maintenance scheduling, they can read Zahedi et al. [4–6].

3.8 Exercises

Try the next cases to understand the method that was developed at this chapter.

1. Let four jobs from single item with job size 25 parts with processing time 500 min. The other parameters of this problem are the same as the problem above. Schedule the jobs with minimization total cost criteria.

2. Let four jobs from single item consist of Job 1 = 40 parts (800 min), Job 2 = 30 parts (600 min), Job 3 = 20 parts (400 min), Job 4 = 10 parts (200 min), and due date d = 3000, and the other parameters of this problem are the same as the problem above. Schedule the jobs with minimization total cost criteria

Author details

Zahedi Zahedi
Mathematics Department, School of Computer Science, Bina Nusantara University, Jakarta, Indonesia

*Address all correspondence to: zahedizahedi@binus.ac.id

References

[1] Dobson G, Karmarkar US, dan Rummel JL. Batching to minimize flow times on one machine. Management Science. 1987;**33**:784-799

[2] Dobson G, Karmarkar US, dan Rummel JL. Batching to minimize flow times on heterogeneous machines. Management Science. 1989;**35**:607-613

[3] Halim AH, Ohta H. Batch scheduling problems to minimize inventory cost in the shop with both receiving and delivery just in times. International Journal of Production Economics. 1994; **33**:185-195

[4] Zahedi Z, Ari Samadhi TMA, Suprayogi S, Halim AH. Integrated batch production and maintenance scheduling for multiple items processed on a deteriorating machine to minimize total production and maintenance costs with due date constraint. International Journal of Industrial Engineering Computations. 2016;**7**(2):229-240

[5] Zahedi R, Yusriski R. Stepwise optimization for model of integrated batch production and maintenance scheduling for single item processed on flow shop with two machines in JIT environment. In: Presented in ICCSCI Bali, 13–15 Oct 2017. Procedia Computer Science. 2017;**116**:408-420

[6] Zahedi Z, Salim A, Yusriski R, Haris H. Optimization of integrated batch production and maintenance scheduling on flow shop with two machines. International Journal of Industrial Engineering Computations (IJIEC). 2019;**10**(2):225-238. DOI: 10.5267/j.ijiec.2018.7.001

[7] Daellenbach HG, McNickle DC. Management Science, Decision Making Through Systems Thinking. New York, USA: Palgrave Macmillan; 2005

[8] Jiang RY, Murthy DNP. Maintenance Decision Models for Management. Beijing: Science Press; 2008

Section 2

Integrated Approach in Process Improvement

Chapter 4

Special Issues of Ensuring Electrical Safety in Networks with Isolated Neutral Voltage up to 1000 V at Mining Enterprises

Bolatbek Utegulov

Abstract

In the practice of operating mining machines and complexes, there are no effective ways to monitor the state of insulation and protect a person from electric shock in a network with voltages up to 1000 V. There is a risk of electric shock with a fatal outcome for personnel. Consequently, the issue of development of methods for monitoring the state of insulation and protection against electric shock in a network up to 1000 V for mining machines and complexes is relevant and urgent. The existing protection of a person from electric shock effectively works provided that the total insulation resistance is commensurate with the capacitive insulation resistance of the phases of the electrical network relative to the ground. But, at mining enterprises, a violation of ratio takes place between the total and capacitive insulation resistance of the network, which leads to failure of the protection against electric shock. Therefore, to ensure electrical safety criteria when operation of electrical installations in the network is up to 1000 V, it is necessary to consider the insulation condition and the technical capabilities of available protection against electric shock in a complex.

Keywords: current, voltage, neutral, insulation, network, resistance

1. Introduction

The implementation of modern technological processes in quarries, which are directly related to the growth of unit capacity of stripping and mining machines, is making increasingly stringent requirements to ensure the safety and security of the operation of electrical system [1, 2].

However, the increase in the length of electrical networks, which feed stripping and mining machines, increases the likelihood of single-phase ground fault, which, as a rule, is the main cause of the interruption of power supply. Relay protection and automation actions allow maintaining the continuity of power supply only if it is possible to control periodic insulation parameters mains phases with respect to ground.

Among the range of issues related to ensuring security of supply of electric power-electrified mining equipment and the safety of its operation, there is a development of methodology for determining the insulation parameters, which occupies a special place, as the results of the method used are derived from the main provisions of organizational and technical measures that promote a culture of service of the internal power supply of mining enterprises. The importance of developing the method to determine the insulation value is also determined by the fact that it can be used in other industries where there is three-phase electrical network with isolated neutral with voltages up to and above 1000 V.

For experimental studies of the state of insulation of three-phase electrical networks with isolated neutral with voltages up to and above 1000 V, a number of methods were proposed [1–9], taking into account the inherent specific characteristics in the internal power supply of open-cast mining. For the insulation parameters measuring methods, a number of requirements are presented, namely:

1. Measurements should be carried out without interruption in the supply of electricity to consumers.

2. The process of measurement should not cause damage to the insulation of electrical networks and electrical accidents.

3. The measurements must be carried out using a small amount of electrical equipment and appliances.

4. Execution of determining the insulation parameters should be safe both for researchers and for personnel servicing electrical systems.

5. Measurements of baseline values should be sufficiently accurate and if possible have a short duration of works on measurement.

6. Accuracy of the method must not exceed 10%.

Based on the analysis of existing methods [1, 2], considering the aforementioned requirements for experimental research of insulation parameters of three-phase electrical networks with isolated neutral, it was concluded that the methods developed earlier are not fully met the essential requirements. Therefore, at present, in mines, previously proposed methods were not used as a primary means of prevention, ensuring uninterrupted power supply and operational safety of electrical installations.

In this regard, there are problems of further improvement of means of controlling insulation parameters of electrical networks in conjunction with the implementation of preventive measures and periodic measurements in different operating conditions. The method of determining the phase insulation parameters to earth of electrical networks must not affect the operation of the electricity system, and the calculation of insulation parameters must contain a minimum of computation.

In practice of operation of electrical networks with isolated neutral with voltage up to 1000 V and above, it is necessary to know the value of the insulation parameters by which the organizational and technical measures are developed to ensure the safety and security of the electricity supply of mining enterprises.

One of the most important issues in the mining industry is the problem of increasing the reliability of power supply systems and reduced level of electrical safety for electrical installations in mines. This condition is associated with physical obsolescence of much of the equipment. Intensive increase in productivity mining leads to complication of the network configuration, which significantly affects the state of electrical networks, reducing the reliability of their operation. At the same time, there is an increase in the number of damages in electrical networks, which are the major causes of wear and the aging network isolation [9].

During coal mining, open-pit mining machines and equipment operate in harsh conditions, which are caused by the constant movement of the front of the mining operations, vibration, dust, and climate-meteorological conditions. This leads to the fact that during operation, the electrical insulation is subjected to a change in the electrical network, changing the properties of the electrical insulating materials. This fact affects the decrease of electrical resistance and electric strength [10].

The main factors of aging of the insulation are operating voltage, transient increase in the voltage at the external and internal overvoltages, oxidation processes caused by the ionization of air and leading to the development of a surface discharge, mechanical effects, bulk and surface contamination, heating, and humidification, influencing the quality of the voltage caused by the use of controlled semiconductor converters. Mining machines' component failure occurs due to changes in the nominal loads; data equipment failures may lead to production downtime [11].

The aforementioned factors intensify the process of reducing the insulation resistance phase of electrical network with respect to earth in coal extraction. Reduced insulation resistance phase electrical network with respect to earth increases the likelihood of the emergency operating modes of the operation of electrical installations, which may be a consequence of electric shocks to persons. With the exception of electric shock, it is necessary to ensure a high level of insulation in a network with an isolated neutral voltage up to 1000 V, through activities related to the systematic and effective control of the condition of insulation. This is one of the main areas to ensure electrical safety in the specific conditions in the development of coal deposits in an open way [12].

According to the "safety regulations for electrical installations," what is required is the mandatory application of the automatic control of insulation with the action off, with periodic measurements of insulation resistance phase of electrical network with respect to earth in electrical installations up to 1000 V [13, 14].

In the development of coal deposits, there is a growing number of electric shocks mainly due to the weak formulation of organizational and technical measures for inspection, repair, and the condition of insulation in electrical networks and electrical equipment. Timely determination of the degree of the deterioration of the insulation can prevent equipment failure [15]. It should be noted that the operational personnel rely on protection against current leakage, which in the operation of electrical networks and electrical equipment may be damaged or artificially out of operation.

Great contributions to the definition of criteria for electrical scientists have been made by the Moscow State Mining Academy, University of California, Georgia State University. One criterion for electrical safety in emergency operation is the limit value of the current flowing through the human body I_h = 6.0 mA, with the contact voltage U_c = 20.0 V with the duration of the current flowing through the human body $t > 1.0$ s, with the main frequency f = 50 Hz, a three-phase network with an isolated neutral voltage up to 1000 V [16].

In general, the analysis of research into condition of insulations and single-phase ground fault current showed that used residual current devices (RCDs) in underground coal mining and mining meet the criteria for electro security under normal and emergency operating modes in a three-phase network with isolated neutral voltages up to 1000 V.

In excavator mining, the electrical network voltage up to 1000 V does not contain lines more than 10 m long, and therefore, the data network is similar to the networks of electric arc furnaces, which are called short. Studies on the condition of insulations in the development of coal deposits and mining open pits in short networks up to 1000 V on excavators are not sufficient to be produced. Installed residual current devices in short networks up to 1000 V excavators have not been studied in relation to the criteria for electric normal and emergency operating modes [17].

Studying the technical parameters of the residual current devices for compliance with electrical safety criteria of normal and emergency operating modes in the three-phase electrical short network with an isolated neutral voltage up to 1000 V is necessary to research the condition of insulation.

The practice of electrical networks up to 1000 V in the development of coal deposits in the enterprise shows a lack of insulation resistance measurement techniques, and if so, then the insulation resistance measurement is made, usually very irregularly with large errors. The most widely used method to measure is found by applying the insulation resistance of the measuring device Megger [1].

It should be noted that the results of the measurements of Megger providing "electric installation code" (EIC) [18] and "rules of technical operation of electrical installations" do not correspond to the real values of the insulation resistance of the network, since the measurements are made in the absence of a working line voltage and disconnected power consumers [19].

Using a Megger measure at low values of insulation resistance in the electrical network and disconnected power consumers allows damage to be established qualitatively. It follows that the use of Megger as a means to assess the conditions of electrical safety for electrical installations is insufficient, since it is impossible to determine the resistance, reactance, and impedance of phase insulation of electrical network with respect to earth under the working voltage [1].

According to the EIC, the rate on the insulation resistance shall not be less than −0.5 MOhm ohmic resistance in the individual circuit element and the electrical network appliance. It is not possible to assess the state of the insulation as a whole. Therefore, the norm EIC relative to −0.5 MW ohmic resistance cannot be accepted as a criterion for operational insulation monitoring conditions and, therefore, as a measure of electrical safety, since from the point of view of safety production work on electrical impedes an evaluation of the insulation and its components [18].

On the basis of the foregoing information, the main task of studying the state of electrical insulation in the development of coal deposits' open method is to determine the main parameters of the insulation of the electrical networks up to 1000 V and to identify factors influencing the state of insulation in conditions of single and bucket wheel excavators and drilling rigs.

Mining companies are currently equipped with high-electrified mining machines that provide high productivity. These companies are heavy consumers of electricity. The power of the electrical installations in the modern excavators reaches 20 MW or more and can be compared with the power of a large industrial company. Indeed, in these companies, reliable and continuous power supply to the electrical receiver depends largely on the condition of the electrical equipment in operation, as well as the intensity of the electrical damage and electrical networks [20].

Frequent movement of flexible cables supplying mobile mining machines leads to mechanical deformation and damage. Thus, attendants are at risk of sustaining electric shocks as they work with cable, electrical equipment, and the metal structure. The number of electrical shocks in an electrical installation is in direct proportion to the frequency of damage to electrical equipment. In this way, more than 80% of electrical shocks are related to direct contact between a man and current-carrying parts, while 3–10% are related to contact between the enclosures of electrical equipment at the time of the existence of single phase-to-earth fault [21].

According to the mine works regulations, safety shutdown is obligatory at mining enterprises. Safety shutdown is fast-operating protection, which automatically switches off the electric equipment under 1000 V when the risk of electric shock is present [22]. This hazard can occur as a result of case-to-phase fault, reduction of phase-to-ground insulation resistance below a certain value, and live-line bare-hand touching [23]. In such cases, residual current devices provide rapid shutdown of the power section. The response time of modern residual current devices (RCDs) does not exceed the time of let-go current supply [24].

To a large extent, the reliability of the electric equipment and the safety of its services depend on the condition of the insulation of live parts of electric equipment [25]. Insulation damage is the major source of accidents and the cause of many electrical shocks with differing levels of severity, as well as fatalities. Insulation monitoring in electrical networks with insulated neutral under 1000 V at mining enterprises is carried out using automatic insulation monitoring devices, such as AIMD-380s, mining protection devices, such as MPD, devices to protect networks from leakage with automatic compensation of capacitive component of leakage current (e.g., PDAC-380), and insulation monitoring devices A-ISOMETER of IRDH575 (Bender) series as well as a number of others.

Automatic insulation monitoring devices are designed to protect people from electric shocks, continuously monitor insulation resistance, and cut off three-phase electric networks with isolated neutral of 50 Hz alternating current in the case of resistance reduction between their phases and earth up to dangerous level. Automatic compensation of the capacitive component of leakage current is used in leakage current protection devices like PDAC, unlike automatic insulation monitoring devices, such as AIMD [26].

On excavators of mining, enterprises use residual current devices such as AIMD, which are designed for mine electric networks, that is, for deep mining. Mine electric networks under 1000 V contain long-distance cable lines, where total admittance of insulation measures is much like capacity admittance of network insulation and active admittance of isolation is lower than total and capacity admittance of isolation. As such, in mine networks, the current in single phase-to-earth fault exceeds the current of the RCD set point. This provides people with effective protection from electric shocks. The effectiveness of RCD in mine electrical networks under 1000 V is shown in the work of professor Manoilov [27].

In the mining industry, it is not uncommon for people to receive electric shocks during maintenance work of excavators and drill-rings when extracting minerals. There are, as yet, no causal inferences of RCD ineffectiveness to protect people from electric shocks during operation of excavators and drill-rings. In order to improve the efficiency of residual current devices, research should be conducted on the condition of insulation in three-phase electric networks with isolated neutral under 1000 V on the excavator.

2. The method of determining the insulation parameters in three-phase electrical networks with isolated neutral with voltages up to and above 1000 V

2.1 Introduction

One of the factors of electric shock is the weakening of insulation condition of a three-phase electrical network with insulated neutral voltages up to and above 1000 V. In order to ensure the increase of efficiency of the power supply system, it is necessary to develop a method of determining the parameters of isolation under operating voltage. Under the effectiveness, we accept ensuring growth of electrical safety and reliability in the operation of electrical installations with voltage up to and above 1000 V. The known [1] method of determining the parameters of isolation, "Ammeter-voltmeter" is a classical method, as it provides a satisfactory accuracy of the unknown quantities, but it does not ensure work safety in electrical installations production works and reduces the reliability of power supply of industrial machinery and equipment. Reduction of electrical installations work reliability and level of electrical safety in the operation of three-phase power networks up to and above 1000 V determined that by using the method "Ammeter-voltmeter," it is necessary to make the metal circuit of a mains phase to earth and measure the total current single-phase fault ground. Since during a metal closure of any phase to earth phase, voltage of the two other phases of the mains with respect to the ground reaches linear values and can thus lead to a short circuit in a multi-phase mains operated, which determines the reliability of power decrease in production machinery. A reduction in electrical safety determined by that in the metal closure of any phase of electrical network and ground, contact voltage, and step voltage will have the maximum value, and thereby provides maximum increase the probability electric shock to persons.

2.2 Method for determining the insulation parameters in an electrical network with insulated neutral

The method presented in the work [6] of determining the insulation parameters in three-phase electrical network with insulated neutral voltages above 1000 V, based on the measurement values of the modules of the line voltage, zero sequence voltage, and phase voltage with respect to ground when connected known active extra conduction between electrical network of the measured phase and ground, has a significant error. A significant error determined by that in determining the insulation parameters using the value of zero sequence voltage module, and thus, it is necessary to use a voltage transformer windings, allowing to allocate the residual voltage.

On the basis of the foregoing methods for determining the insulation parameters in three-phase mains with insulated neutral voltages up to and above 1000 V, which provides a satisfactory accuracy of the unknown quantities by eliminating the measurement of the modulus of the residual voltage, the operational safety of electrical installations, and the reliability of the electricity system, in connection excluding the measurements of the total current of the module for single-phase earth fault between a mains phase with respect to ground.

A method for determining the insulation parameters in three-phase balanced networks with voltage up to and above 1000 V, based on the measurement values of the modules of the line voltage, the phase voltages A and C relative to the ground after connecting additional active conductivity between the phase A and the mains ground was developed.

As a result of the measurement values of the modules of the line voltage and phase voltage C and A with respect to the ground, taking into account the magnitude of the additional active conductivity by mathematical formulas, the following are defined:

- the total conductance of network insulation

$$y = \frac{1.73U_1U_A}{U_C^2\text{-}U_A^2}g_o, \tag{1}$$

- the active conductance of network insulation

$$g = \left(\frac{3U_l^2(U_1^2 - 3U_A^2)}{\left(U_C^2 - U_A^2\right)^2} - 1\right)0.5g_o, \tag{2}$$

- capacitive conductance of network insulation

$$b = \left(y^2 - g^2\right)^{0.5}, \tag{3}$$

where U_l is the line voltage; U_A is the A phase voltage with respect to the ground; U_C is C the phase voltage with respect to the ground; and g_o is the additional active conductance.

The method developed in the implementation does not require the creation of a special measuring device, since the measuring devices, that is, voltmeters, available in the service manual. The PE-200 resistance is used as an active additional conductivity with R = 1000 Ohms, where by means of parallel and serial connection provides the required power dissipation. To switch, the active standby is used more conductivity cell load switch.

The developed method provides satisfactory accuracy and is simple and safe in its implementation in the three-phase electrical networks with isolated neutral voltages up to and above 1000 V.

2.3 Analysis of error of method determining the insulation parameters in an electrical network with isolated neutral

The obtained mathematical dependences for determining the total and active conductance of electrical network insulation provide easy and safe work of electrical installations with voltage up to and above 1000 V.

Error analysis of the developed method for determining the insulation parameters in symmetrical three-phase electrical networks with isolated neutral which is based on measurement of unit line voltage, phase voltage C and A relative to the earth, after the active connection of additional conduction between phase A and the electric network and earth is performed.

To improve the efficiency of the developed method for determining the parameters of isolation in a symmetrical three-phase network with isolated neutral, based on error analysis, for each specific network, additional active conductivity is selected, in order to ensure satisfactory accuracy of required quantities.

Random relative error in determining the total conductivity of insulation and its components in three-phase balanced networks with voltage up to and beyond 1000, based on the measurement values of the modules of the line voltage, phase voltage C and A with respect to the ground, after connecting the active additional conduction between the phase and the electric network and earth, is determined according to (1), (2), and (3).

Random relative error in determining the total conductance of mains phase insulation relative to the ground is determined from the formula (1):

$$y = \frac{1.73U_1U_A}{U_C^2\text{-}U_A^2} g_o,$$

where U_l, U_A, U_C, and g_o are values that define the total conductance of network insulation and obtained by direct measurement. The relative mean square error in determining the total conductance of mains phase insulation relative to the ground is determined from the expression [28, 29]:

$$\Delta y = \frac{1}{y}\left[\left(\frac{\partial y}{\partial U_A}\Delta U_A\right)^2 + \left(\frac{\partial y}{\partial U_C}\Delta U_C\right)^2 + \left(\frac{\partial y}{\partial U_l}\Delta U_l\right)^2 + \left(\frac{\partial y}{\partial g_o}\Delta g_o\right)^2\right]^{0.5}, \quad (4)$$

where $\frac{\partial y}{\partial U_A}$, $\frac{\partial y}{\partial U_C}$, $\frac{\partial y}{\partial U_l}$, and $\frac{\partial y}{\partial g_o}$ are partial derivatives $y = f\ (U_l, U_A, U_C, g_o)$.

Here ΔU_l, ΔU_A, ΔU_C, and Δg_o are absolute errors of direct measurement values U_l, U_A, U_C, and g_o which are defined by the following expressions:

$$\begin{aligned} \Delta U_l &= U_l \times \Delta U_{l^*}; \\ \Delta U_C &= U_C \times \Delta U_{C^*}; \\ \Delta U_A &= U_A \times \Delta U_{A^*}; \\ \Delta g_o &= g_o \times \Delta g_{o*}. \end{aligned} \quad (5)$$

To determine the errors of measuring devices, accept that $\Delta U_{l*} = \Delta U_{A*} = \Delta U_{C*} = \Delta U_*$, where: ΔU_* is the relative error of voltage measurement circuits and $\Delta g_{o*} = \Delta R_*$ is the relative error of the measuring instrument, which measures the resistance which is connected between the phase A electrical and ground. Determine the partial derivative functions $y = f\ (U_1, U_A, U_C, g_o)$ by the variables U_1, U_A, U_C, g_o:

$$\begin{aligned} \frac{\partial y}{\partial U_1} &= \frac{1.73U_A}{U_C^2 - U_A^2} g_o; \\ \frac{\partial y}{\partial U_A} &= \frac{1.73U_1(U_C^2 + U_A^2)}{(U_C^2 - U_A^2)^2} g_o; \\ \frac{\partial y}{\partial U_C} &= -\frac{3.46U_1U_AU_C}{(U_C^2 - U_A^2)^2} g_o; \\ \frac{\partial y}{\partial g_o} &= \frac{1.73U_1U_A}{U_C^2 - U_A^2}. \end{aligned} \quad (6)$$

Solving the Eq. (4), substituting the values of the partial derivatives of Eq. (6) and private values of absolute errors (5), at the same time, assuming that $\Delta U_* = \Delta R_* = \Delta$, we obtain:

$$\varepsilon_y = \frac{\Delta y}{\Delta} = \frac{1.73U_lU_Ag_o}{U_C^2 - U_A^2}\left(2 + \frac{4U_C^4 + (U_C^2 + U_A^2)^2}{(U_C^2 - U_A^2)^2}\right)^{0,5}. \quad (7)$$

The obtained Eq. (7) is divided into the Eq. (1):

$$\varepsilon_y = \frac{\Delta y}{\Delta} = \left(2 + \frac{4U_C^4 + \left(U_C^2 + U_A^2\right)^2}{\left(U_C^2 - U_A^2\right)^2}\right)^{0,5} \tag{8}$$

The obtained Eq. (8) is expressed in relative units, and after the conversion, we obtain:

$$\varepsilon_y = \frac{\Delta y}{\Delta} = \left(2 + \frac{4 + \left(1 + U_*^2\right)^2}{\left(1 - U_*^2\right)^2}\right)^{0,5}, \tag{9}$$

where $U_* = \frac{U_A}{U_C}$.

Random error in determining the active conductance of mains phase insulation relative to the ground is determined from the formula (2):

$$g = \left(\frac{3U_l^2\left(U_l^2 - 3U_A^2\right)}{\left(U_C^2 - U_A^2\right)^2} - 1\right)0.5g_o,$$

where U_l, U_A, U_C, and g_o are values that define the active conductance of network isolation and obtained by direct measurement.

Relative mean square error of the method when determining the active conductivity of phase insulation of electrical network relative to the ground is determined from the expression:

$$\Delta g = \frac{1}{g}\left[\left(\frac{\partial g}{\partial U_A}\Delta U_A\right)^2 + \left(\frac{\partial g}{\partial U_C}\Delta U_C\right)^2 + \left(\frac{\partial g}{\partial U_l}\Delta U_l\right)^2 + \left(\frac{\partial g}{\partial g_o}\Delta g_o\right)^2\right]^{0.5}, \tag{10}$$

where $\frac{\partial g}{\partial U_A}$, $\frac{\partial g}{\partial U_C}$, $\frac{\partial g}{\partial U_l}$, and $\frac{\partial g}{\partial g_o}$ are partial derivatives, $g = f\ (U_l, U_A, U_C, g_o)$.

Here ΔU_l, ΔU_A, ΔU_C, and Δg_o are absolute errors of direct measurement values U_l, U_A, U_C, and g_o, which are defined by the following expressions:

$$\begin{aligned} \Delta U_l &= U_l \cdot \Delta U_{l*}; \\ \Delta U_C &= U_C \cdot \Delta U_{C*}; \\ \Delta U_A &= U_A \cdot \Delta U_{A*}; \\ \Delta g_o &= g_o \cdot \Delta g_{o*}. \end{aligned} \tag{11}$$

To determine the accuracy of measuring devices, accept that $\Delta U_{l*} = \Delta U_{A*} = \Delta U_{C*} = \Delta U_*$, where ΔU_* is the relative error of voltage measurement circuits and $\Delta g_{o*} = \Delta R_*$ is the relative error of a measuring instrument that measures resistance which is connected between the phase A electrical and the ground.

Determine the partial derivatives $g = f\ (U_l, U_A, U_C, g_o)$ by the variables U_l, U_A, U_C, and g_o:

$$\begin{aligned} \frac{\partial g}{\partial U_l} &= \frac{3U_l\left(2U_l^2 - 3U_A^2\right)}{2\left(U_C^2 - U_A^2\right)^2}g_o; \\ \frac{\partial g}{\partial U_A} &= -\frac{3U_l^2 U_A\left(3U_C^2 + 3U_A^2 - 2U_l^2\right)}{\left(U_C^2 - U_A^2\right)^3}g_o; \\ \frac{\partial g}{\partial U_C} &= -\frac{6U_l^2 U_C\left(U_l^2 - 3U_A^2\right)}{\left(U_C^2 - U_A^2\right)^3}g_o; \\ \frac{\partial g}{\partial g_o} &= \frac{3U_l^2\left(U_l^2 - 3U_A^2\right)}{2\left(U_C^2 - U_A^2\right)} - 0.5. \end{aligned} \tag{12}$$

Solve Eq. (10), substituting the values of the partial derivatives of Eq. (12) and the values of the partial absolute errors (11), at the same time, assuming that $\Delta U_* = \Delta R_* = \Delta$, we obtain:

$$\frac{\Delta g}{\Delta} = \frac{3g_o}{\left(U_C^2 - U_A^2\right)^3}\left(\begin{matrix}\left(U_C^2 - U_A^2\right)^2\left[2U_l^4\left(U_l^2 - 3U_A^2\right)^2 - \left(U_C^2 - U_A^2\right)^4\right] + \\ + U_l^4\left\{U_A^4\left[3\left(U_C^2 - U_A^2\right) - 2U_l^2\right]^2 + U_C^4\left(U_l^2 - 3U_A^2\right)^2\right\}\end{matrix}\right)^{0.5} \tag{13}$$

Obtained Eq. (13) divided by Eq. (2):

$$\varepsilon_g = \frac{\Delta g}{\Delta} = \left(\begin{matrix}\frac{2U_l^4\left(U_l^2 - 3U_A^2\right)^2 - \left(U_C^2 - U_A^2\right)^4}{\left[3U_l^2\left(U_l^2 - 3U_A^2\right) - \left(U_C^2 - U_A^2\right)^2\right]^2} + \\ + \frac{U_l^4\left\{U_A^4\left[3\left(U_C^2 - U_A^2\right) - 2U_l^2\right]^2 + U_C^4\left(U_l^2 - 3U_A^2\right)^2\right\}}{\left(U_C^2 - U_A^2\right)^2\left[3U_l^2\left(U_l^2 - 3U_A^2\right) - \left(U_C^2 - U_A^2\right)^2\right]^2}\end{matrix}\right)^{0.5}. \tag{14}$$

In the resulting Eq. (14), the value of the line voltage is expressed in terms of the phase voltages in accordance with the fact that $U_l = 1.73U_\phi$:

$$\varepsilon_g = \frac{\Delta g}{\Delta} = 3\left(\begin{matrix}\frac{18U_{ph}^4\left(U_{ph}^2 - U_A^2\right)^2 - \left(U_C^2 - U_A^2\right)^4}{\left[27U_{ph}^2\left(U_{ph}^2 - U_A^2\right) - \left(U_C^2 - U_A^2\right)^2\right]^2} + \\ + \frac{3U_{ph}^4U_A^4\left(U_C^2 - U_A^2 - 2U_{ph}^2\right)^2 + U_C^4\left(U_{ph}^2 - U_A^2\right)^2}{\left(U_C^2 - U_A^2\right)^2\left[27U_{ph}^2\left(U_{ph}^2 - U_A^2\right) - \left(U_C^2 - U_A^2\right)^2\right]^2}\end{matrix}\right)^{0.5} \tag{15}$$

Simplifying the formula (15), we obtain the Eq. (16):

$$\varepsilon_g = \frac{3}{27U_{ph}^2\left(U_{ph}^2 - U_A^2\right) - \left(U_C^2 - U_A^2\right)^2}\left(\begin{matrix}18U_{ph}^4\left(U_{ph}^2 - U_A^2\right)^2 - \left(U_C^2 - U_A^2\right)^4 + \\ + \frac{3U_{ph}^4U_A^4\left(U_C^2 - U_A^2 - 2U_{ph}^2\right)^2}{\left(U_C^2 - U_A^2\right)^2} + \\ + \frac{U_C^4\left(U_{ph}^2 - U_A^2\right)^2}{\left(U_C^2 - U_A^2\right)^2}\end{matrix}\right)^{0.5} \tag{16}$$

Obtained Eq. (16) is expressed in relative units and after the conversion, we obtain:

$$\varepsilon_g = \frac{\Delta g}{\Delta} = \frac{3}{27\left(1 - U_{A*}^2\right) - \left(U_{C*}^2 - U_{A*}^2\right)^2}\left(\begin{matrix}18\left(1 - U_{A*}^2\right)^2 - \left(U_{C*}^2 - U_{A*}^2\right)^4 + \\ + \frac{3U_{A*}^4\left(U_{C*}^2 - U_{A*}^2 - 2\right)^2}{\left(U_{C*}^2 - U_{A*}^2\right)^2} + \\ + \frac{U_{C*}^4\left(1 - U_{A*}^2\right)^2}{\left(U_C^2 - U_A^2\right)^2}\end{matrix}\right)^{0.5}, \tag{17}$$

where $U_{A*} = \frac{U_A}{U_{ph}}$ and $U_{C*} = \frac{U_C}{U_{ph}}$.

Relative mean square error method for determining the conductivity of the capacitive isolation mains phases relative to the ground is determined by the expression (3):

$$\Delta b = \frac{1}{b}\left[\left(\frac{\partial b}{\partial y}\Delta y\right)^2 + \left(\frac{\partial b}{\partial g}\Delta g\right)^2\right]^{0.5}, \tag{18}$$

or

$$\varepsilon_b = \frac{\Delta b}{\Delta} = \frac{\left[(1-\tan^2\delta)^2\left(\frac{\Delta y}{\Delta}\right)^2 + \left(\frac{\Delta g}{\Delta}\right)^2\right]^{0.5}}{\tan^2\delta}. \tag{19}$$

Solving Eq. (19) and substituting the values of mathematical descriptions of the relative rms dependences of total (8) and active (16) conductivities of electrical installations phase insulation relative to the ground phase, we get the following equation:

$$\varepsilon_b = \frac{\Delta b}{\Delta} = \frac{\left(\begin{array}{l}(1-\tan^2\delta)^2\left[2+\frac{4U_C^4+\left(U_C^2+U_A^2\right)^2}{\left(U_C^2-U_A^2\right)^2}\right]+ \\ +\frac{9}{\left[27U_{ph}^2\left(U_{\mathrm{ph}}^2-U_A^2\right)-\left(U_C^2-U_A^2\right)^2\right]^2}\times \\ \times\left[\begin{array}{l}18U_{ph}^2\left(U_{\mathrm{ph}}^2-U_A^2\right)^2-(U_C^2-U_A^2)^4+ \\ +\frac{3U_{ph}^4U_A^4\left(U_C^2-U_A^2-2U_{ph}^2\right)^2+U_C^4\left(U_{\mathrm{ph}}^2-U_A^2\right)^2}{\left(U_C^2-U_A^2\right)^2}\end{array}\right]\end{array}\right)^{0.5}}{\tan^2\delta}. \tag{20}$$

Obtained Eq. (21) is expressed in relative units and after the conversion, we obtain:

$$\varepsilon_b = \frac{\Delta b}{\Delta} = \frac{\left(\begin{array}{l}(1-\tan^2\delta)^2\left[2+\frac{4U_{C*}^4+\left(U_{C*}^2+U_{A*}^2\right)^2}{\left(U_{C*}^2-U_{A*}^2\right)^2}\right]+ \\ +\frac{9}{\left[27\left(1-U_{A*}^2\right)-\left(U_{C*}^2-U_{A*}^2\right)^2\right]^2}\times \\ \times\left[\begin{array}{l}18\left(1-U_{A*}^2\right)^2-\left(U_{C*}^2-U_{A*}^2\right)^4+ \\ +\frac{3U_{A*}^4\left(U_{C*}^2-U_{A*}^2-2\right)^2+U_{C*}^4\left(1-U_{A*}^2\right)^2}{\left(U_{C*}^2-U_{A*}^2\right)^2}\end{array}\right]\end{array}\right)^{0.5}}{\tan^2\delta}. \tag{21}$$

Based on the results of random relative mean square errors in determining the active, capacitive, and total conductivities of mains phase isolation relative to the ground, build the dependence:

$$\varepsilon_y = \frac{\Delta y_*}{\Delta} = f(U_*);$$

$$\varepsilon_g = \frac{\Delta g_*}{\Delta} = f(U_{A*}, U_{C*});$$

$$\varepsilon_b = \frac{\Delta b_*}{\Delta} = f(U_{A*}, U_{C*}, \tan\delta),$$

shown in **Figures 1–3**. Mathematical dependence of the relative mean square errors of the total—ε_y, active—ε_g, and capacitive—ε_b conductivities of phase insulation of electrical network with insulated neutral on graphic illustrations

(**Figures 1–3**) characterize the change in error depending on the amount of additional active conduction g_o, which is inserted between the A-phase of electrical network and earth.

In determining the parameters of isolation in a symmetrical three-phase electrical network with isolated neutral on the basis of the method of analysis of error for each specific network, select additional active conduction, so as to ensure the satisfactory accuracy required.

In determining the total conductance of mains phases isolation relative to the ground is chosen such additional active conductivity, the values were within $U_* =$ 0.2–0.8, at the same time as shown in **Figure 1**, the error does not exceed 5% when using measuring devices with accuracy class 1.0, and 2.5% when using measuring devices with accuracy class 0.5.

In determining the value of the active conductance in the three-phase electrical network with insulated neutral voltage up to 1000 V and above, select this additional g_o, so that U_{A*} = 0.2–0.8, when U_{C*} = 1.1–1.6, then on the basis of graphic illustrations of **Figure 2**, error does not exceed 3.5% when using measuring devices with accuracy class 1.0.

In determining the capacitive conductance mains phase isolation relative to the ground selection of additional active conductance g_o based on a graphic illustrations of **Figure 3** so that U_{A*} = 0.2–0.8, when U_{C*} = 1.1–1.6, when tan δ = 1.0, to provide error to 4% when using measuring devices with accuracy class 1.0.

It should be noted that when using measuring instruments with an accuracy class of 0.5, errors of ε_y—total, ε_g—active, ε_b—capacitive admittances of isolation is reduced by half, to provide more reliable data when determining the insulation parameters developed method.

According to the research undertaken by Professor L. Gladilin, a method was developed for determining the parameters of the insulation in networks with an isolated neutral voltage up to 1000 V (method ammeter-voltmeter) [1]. The disadvantage of the method ammeter-voltmeter is the production of single-phase ground

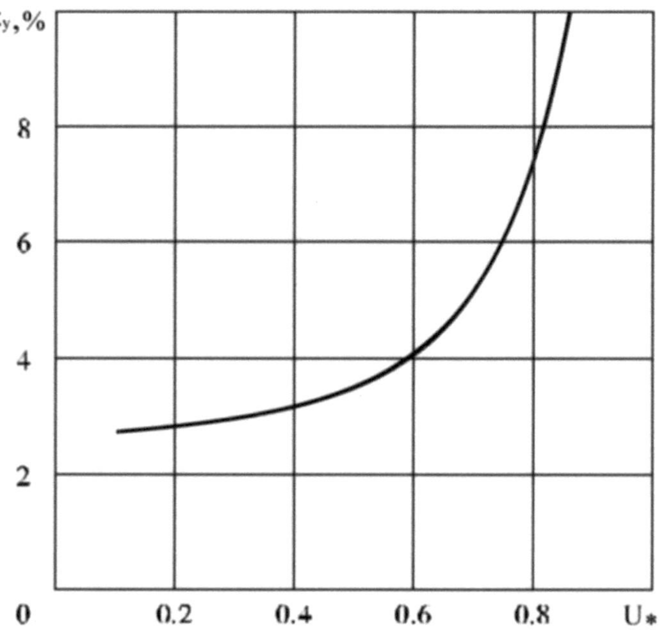

Figure 1.
Analysis of the error in determining the total conductance of the network insulation.

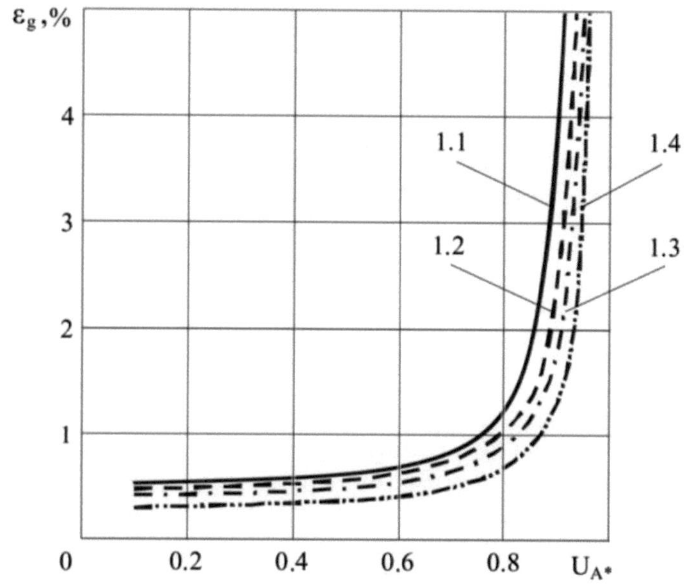

Figure 2.
Analysis of the error in determining the active conductance of the network insulation. U_{C*} = 1.1; 1.2; 1.3; 1.4.

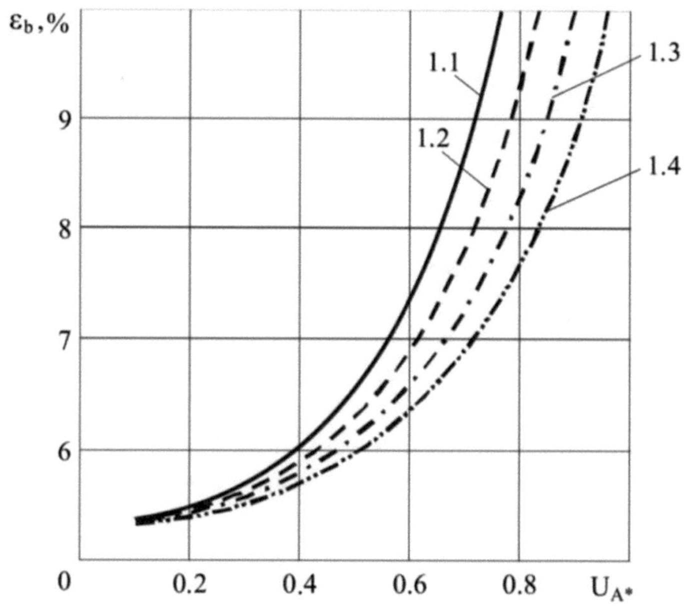

Figure 3.
Analysis of the error in determining the capacitive conductance of the network insulation when tan δ = 1.0. U_{C*} = 1.1; 1.2; 1.3; 1.4.

fault current measurement in the study of a three-phase power network with an isolated neutral. When measuring single-phase ground fault current in three-phase power network, the magnitude-phase voltage is equal to zero. The voltages of the other two phases achieve linear value, it can lead to a two- or three-phase short circuit, and it is emergency operating mode. This leads to a break in supply, as well as increased contact voltage, which is dangerous in the operation of mining machines and systems [1].

The developed method provides satisfactory accuracy when determining the parameters of isolation, as well as the ease and safety of production work in existing electrical installations voltages up to and above 1000 V.

3. Modeling method for measuring the admittance of insulation in a network with an isolated neutral voltage up to 1000 V in mines using Matlab/Simulink

3.1 Introduction

Note that the conductance characterizes the insulating properties of the dielectric, and the susceptance, respectively, characterizes the network capacity, that is, the number of connected electrical receivers and the length of overhead lines and cables. Admittance characterizes the single-phase ground fault current. Therefore, in practice, it is necessary to know the operation of the electrical conductance, susceptance, and admittance of phase of electrical network with respect to earth. This will allow choosing the right strategy to develop organizational and technical measures to increase the level of electrical networks up to 1000 V in the development of coal deposits [30].

Developed in [31], a phase-sensitive method for determining the parameters of insulation in a symmetric network with an isolated neutral voltage up to 1000 is based on the measurement of the modulus of the line voltage and phase voltage to earth after the connection between it and the earth an additional conductance and measuring the phase angle between the vector of the line voltage and vector of the phase voltage to earth. The above phase-sensitive method for determining the insulation contains significant disadvantages in using a special measuring device for measuring the phase angle between the voltage vectors.

3.2 Theoretical studies of the insulation on the basis of a circular chart

For simplicity of measurements, consider a method for measuring the admittance of insulation in a network with an isolated neutral voltage up to 1000 V [7]:

$$y = \frac{U_{\text{ph}o}}{U_o} g_o, \tag{22}$$

where $U_{\text{ph}o}$ is phase voltage to earth after connecting additional conductance g_o; g_o is additional conductance; and U_o is zero phase-sequence voltage.

To measure the admittance of insulation in a network in accordance with the formula (22), it is necessary to enter into the electrical network adjustable resistance between the phase of network and earth. Changing the value of resistance between the phase of network and earth will change the quantities of module phase voltage to earth and zero phase-sequence voltage. From Eq. (22) is obtained the conclusion that with equal admittance of insulation in a network and additional conductance, which is inserted between the phase of network and earth, measured values of the quantities of module phase voltage to earth and zero phase-sequence voltage will be equal:

$$y = g_o \text{ at } U_{\text{ph}o} = U_o.$$

On the basis of the foregoing information, for measuring the admittance of insulation in a network with an isolated neutral, it is necessary to enter adjustable resistance to fulfill equality conditions between the values of the modules phase voltage to earth and zero phase-sequence voltage $U_{\text{ph}o} = U_o$.

To determine the conductance network isolation by using the equal quantities of module phase voltage to earth and zero phase-sequence voltage $U_{pho} = U_o$, equation is

$$y = \frac{U_{ph}^2 - 2U_{pho}^2}{2U_{pho}^2} g_o. \quad (23)$$

Capacitive susceptance of isolation is found as a geometric difference between the admittance of insulation and conductance [7].

The method for determining the parameters of insulation in a network with an isolated neutral voltage above 1000 V describes circular chart changes in the modulus phase voltage to earth and zero phase-sequence voltage as shown in **Figure 4**. Changes in the modulus phase voltage to earth and zero phase-sequence voltage are produced in accordance with the circular chart of changes in the magnitude of the additional conductance.

Figure 4 shows the phase voltages U_{ph} of three phases A, B, and C, before connecting additional conductance to phase A; neutral-point displacement voltage U_o; and phase voltage to earth after connecting additional conductance g_o to phase A—U_{pho}. Point 0_2 corresponds to equal quantities of U_{pho} and U_{o2}.

Experimental studies of the circular chart have shown that changes of the modulus phase voltage to earth and zero phase-sequence voltage depends on the selection of the magnitude of the additional conductance. Hereby it is consistent with the fundamental provisions of the theoretical fundamentals of electrical engineering.

3.3 The method of measuring the admittance in a network with an isolated neutral voltage up to 1000 V

To ensure the equal quantities of phase voltage to earth and zero phase-sequence voltage with the connected additional conductance, an additional conductance

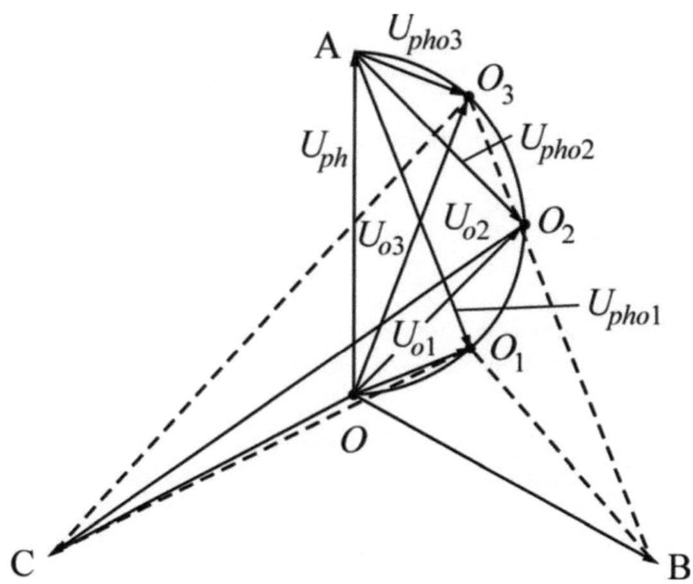

Figure 4.
A circular chart changes the modulus phase voltage to earth and zero phase-sequence voltage, depending on the size of the additional conductance.

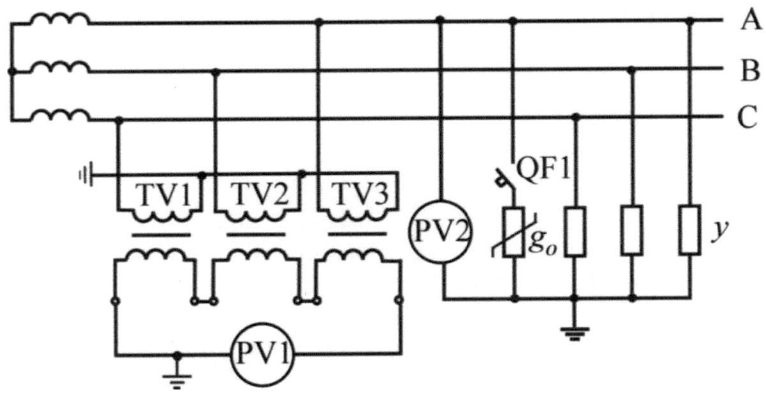

Figure 5.
Electrical schematic diagram of a method of measuring the admittance of network.

variable resistance is used. The variable resistor is connected between the measured electrical network phase and earth. Then the resistance regulation is provided to ensure equality between the voltage phase to earth and zero sequence voltage. In case of equal voltage, magnitude of admittance will correspond to the value of variable resistance, which is connected between the phase of network and earth. The method for measuring the admittance of insulation in a network with an isolated neutral voltage up to 1000 V will provide improved accuracy and speed measurement admittance network insulation [32].

The measurements of phase voltage to earth and zero phase-sequence voltage produced an AC voltmeter. The zero phase-sequence voltage is released from the network by using three single-phase transformers; the primary windings are connected in a star and the secondary windings are connected into an open triangle.

Developing the method of measuring the admittance in a network with an isolated neutral voltage 1000 V is explained in the schematic circuit diagram shown in **Figure 5**. The electrical schematic circuit comprises the electrical network, with phases *A*, *B*, and *C*; three single-phase voltage transformers *TV1*, *TV2*, and *TV3*; voltmeter *PV1*, measured quantities of module of the zero phase-sequence voltage; voltmeter *PV2*, measured module of phase voltage to earth; switching device *QF1*, introduction of adjustable additional conductance; additional conductance *go*; and admittance of network *y*.

The method is as follows: for measuring the admittance of the network, the voltmeter *PV2* measures the phase voltage to earth; the voltmeter *PV1* measures the zero phase-sequence voltage on the secondary winding of single-phase voltage transformers *TV1*, *TV2*, and *TV3*. Switching device *QF1* connects the adjustable additional conductance *go*, making the regulation of the magnitude of additional conductance to achieve the equality of the modulus of the phase voltage to earth and the zero phase-sequence voltage. In this case, the value of additional conductance will fit the admittance of network [32].

3.4 Modeling method of measuring the admittance of insulation in a network with an isolated neutral voltage up to 1000 V using Matlab/Simulink

As a tool for analyzing the operating conditions of power network, the package Matlab/Simulink has been used. The package has a sufficiently developed set of special blocks for modeling elements of the power system.

Matlab/Simulink enables an electrical schematic diagram of a method of measuring the admittance of insulation to be implemented (**Figure 6**). The diagram

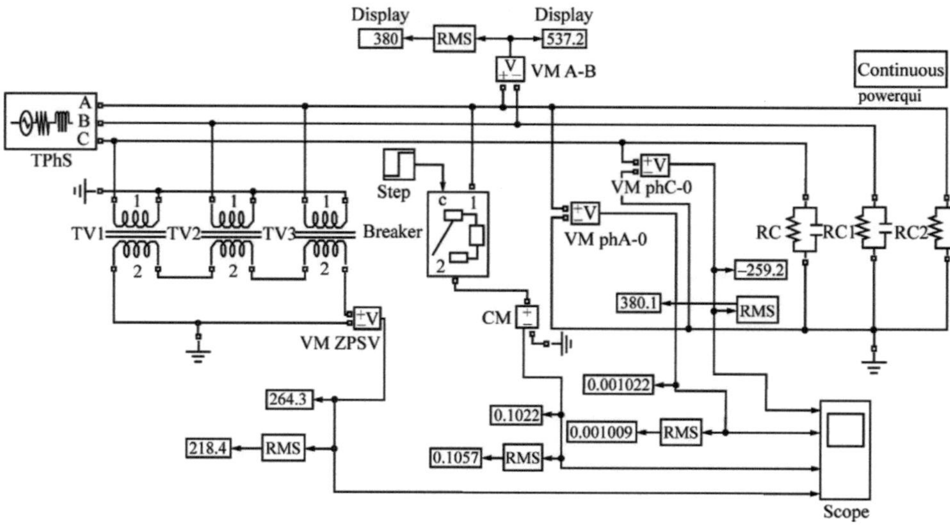

Figure 6.
Simulation model of the metal single phase ground short circuit.

comprises three-phase source voltage 380/220 V; active-reactive resistance RC-RC2; zero phase-sequence voltage filter made by three single-phase voltage transformers; voltage and current measurements; oscillograph (scope) to display network settings; and displays to show the amplitude and true values of the electrical parameters.

To test the electrical schematic diagram, operating conditions were simulated by using a metal single-phase ground short circuit, which was carried out by way of a block breaker. A time-modulating circuit of 0.2 s was implemented by way of a block step. Block RMS allows for the calculation of the true RMS value of the input signal [33].

In the metal single-phase A ground short circuit operating conditions, the amplitude value of phase C voltage to earth was equal to 537 V, which corresponds to the true value of 380 V. The RMS value of phase A voltage to earth after the metal single-phase ground short circuit corresponds to 0.001009 V, with the current value of the faulty phase A being equal to 0.1057 A. From the findings of the zero phase-sequence voltage filter, the true value of the voltage increased from zero to 218.4 V (**Figure 7**).

The simulation model of the method of measuring the admittance of insulation with variable resistor R is shown in **Figure 8**. From the diagram, as a variable resistor R is connected between the measured phase of network and earth, a nonlinear resistor R diagram is used [34].

An application of the variable resistor enables the production of multiple controls for the electrical network parameters. According to the method described previously, a switching device is introduced to adjust the additional conductance. A block breaker is then added to the switching device. The additional conductance is represented by a variable resistor, namely, nonlinear R. A subsystem of the variable resistor R is shown in **Figure 9**. It is possible to use block slider gain to adjust the parameters of the resistor.

In the diagram, the controlled current source is connected in parallel with a voltage measurement. Between the output of the voltage measurement and the input of the controlled current source, the Simulink model is turned on, which implements the voltage-current characteristic of the device. In parallel to the controlled current source, decoupling resistor series RLC branch is also connected. Its

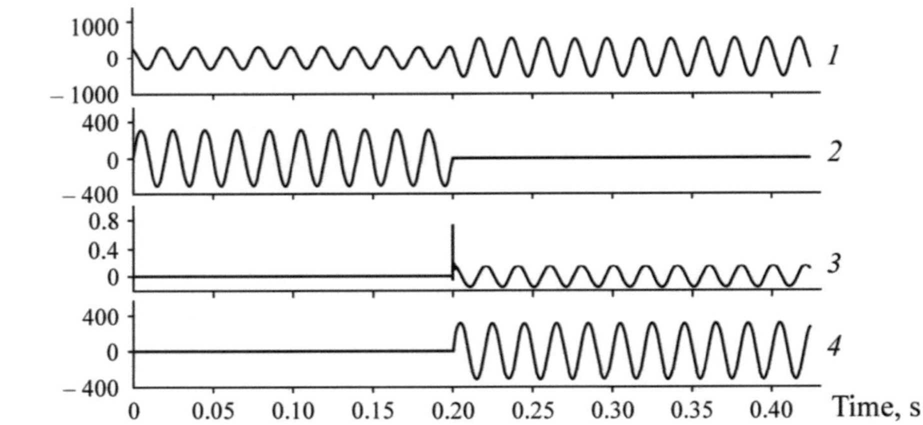

Figure 7.
Voltage and current for metal single phase ground short circuit: 1—phase C voltage to earth, V; 2—phase A voltage to earth, V; 3—current under A phase-to-ground fault, A; 4—zero phase-sequence-voltage, V. Phase-to-ground fault time 0.2 s.

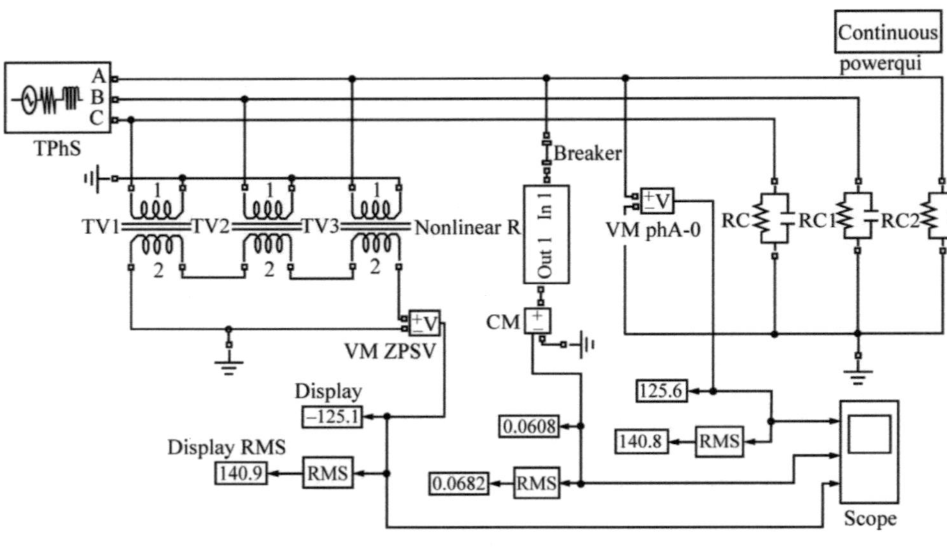

Figure 8.
Simulation model of a method of measuring the admittance of insulation in Simulink: display RMS—display of the actual data of power grid; nonlinear R—variable resistor to adjust extra conductance, the rest of legend find in ***Figure 3****.*

presence is due to the fact that a large number of SimPowerSystems blocks are made on the basis of the current sources. When these blocks are connected in series, the current sources are also connected in series which is unacceptable. The presence of the decoupling resistor enables the connection of these blocks in series. The value of the resistor chosen should be sufficiently large to minimize its effect on the characteristics of the created block [34].

The Simulink model of the variable resistor is implemented using a block slider gain, which allows for a change in scalar gain during the simulation using the slider. Thus, the value of the slider gain is regulated until the zero phase-sequence voltage equals the *A* phase voltage to earth.

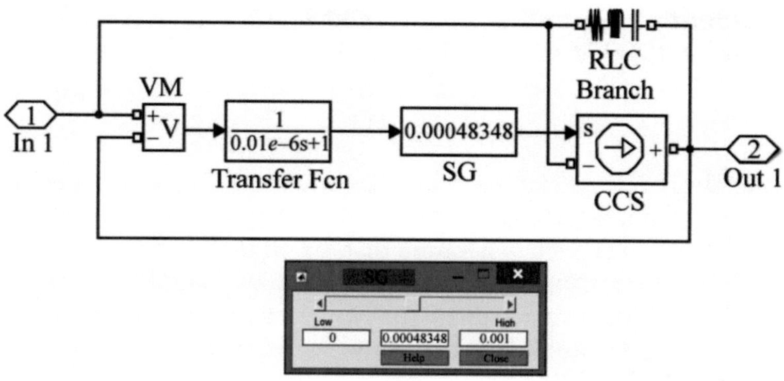

Figure 9.
Subsystem of the variable resistor R: VM—voltage meter; transfer Fcn—gain transfer characteristic block; SG—resistor adjustment slide; RLC branch—decoupling resistor; CCS—adjustable current source.

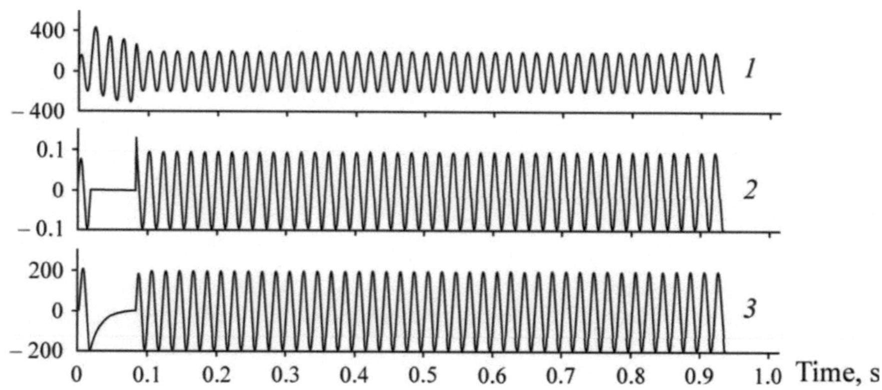

Figure 10.
Phase voltage, current and zero phase-sequence voltage: 1—phase A voltage to earth, V; 2—phase A current to earth, A; 3—zero phase-sequence-voltage, V.

Due to the data received in the regulation of the variable resistor to 2068 Ohms, the true value of the zero phase-sequence voltage and phase A voltage to earth are equal to 140.8 V. **Figure 10** shows the amplitude values of the phase voltage and current and zero phase-sequence voltage for the value of the variable resistor R = 2068 Ohms.

Thus, according to the circular chart above, when phase voltage to earth U_A equals the zero phase-sequence voltage U_o, the variable of the admittance of insulation y corresponds to the variable resistance which is connected between A phase and earth.

According to the developed method of measuring the admittance of isolation in a network with isolated neutral voltage up to 1000 V, the variable of the admittance y corresponds to 2068 Ohms which composes 0.48 mS.

The simulation model of the method of measuring the admittance of insulation in the Matlab/Simulink environment allows for the regulation of variable resistor to be used and to simplify the calculations of the magnitude of the admittance of insulation in a network with isolated neutral voltages up to 1000 V. The method for measuring the admittance of insulation in a network with an isolated neutral voltage up to 1000 V will allow for an increase in the accuracy and speed of measurement of the admittance of network.

4. Development of method to improve efficiency of residual current device under 1000 V on excavators of mining enterprises

4.1 Introduction

The principle of voltage stabilization in the system with the series SC and variable frequency can be explained with a vector diagram of the first harmonics of current and voltages. For such generators that are on the basis of the numerical values of the insulation parameters, it is clear that loss-angle tangent and current in single phase-to-earth fault can evaluate the work of a residual current device on the excavator. Experimental studies regarding the parameters of insulation, loss-angle tangent, and current in single phase-to-earth fault have evaluated the condition of short network under 1000 V in terms of electrical production work in the operation of electrical equipment excavators [1].

The experimental studies were carried out in the coal mine Ekibastuz, Angrensor LLP in the Pavlodar Region of Kazakhstan, to establish the actual values of the basic parameters of the insulation of electrical networks under 1000 V on the excavator EKG-8I.

Compulsory use of the system with insulated neutral networks under 1000 V on the excavators is caused by electricity safety conditions. During the process of solving electrical safety issues in the mining industry, considerable expertise has been built up, especially in the field of research on the condition of electrical insulation with insulated neutral voltage under 1000 V [1]. Neither the methods used to investigate the insulation condition nor the results of these studies can be taken for excavators because the power supply system of the excavator has its own characteristics, namely that there are no long cable lines and the electrical receivers are concentrated in a small area. Put differently, the power of the electrical receiver, such as excavator EKG-8I, is produced by a short network. Electrical receivers operate in different geological, climatic, and meteorological conditions, and this also affects the measurements.

4.2 The study of the condition of insulation

In surveyed excavators, with Ekibastuz coal mine used as the insulation control device, safety rules were prescribed, while automatic insulation monitoring devices were applied, including AIMD, and leakage current LC-2M. Experience in operating electrical equipment positively recommended leakage relay, such as AIMD, which are designed to mine district networks with voltage under 1000 V, that is, for deep mining [14]. Mine networks with voltage under 1000 V contain lengthy branching cable lines, which are powered by the electrical receivers of mining machines and systems. The main reason of ineffective relay is discrepancy between the technical capabilities of RCD and parameters of the insulation network voltage under 1000 V on excavator. However, in most cases, the existing power supply for excavators, which have, as a rule, one main substation, does not make it possible to meet this requirement. This is due to the fact that the relay AIMD, which is a device providing network-wide protection from leakage, switches off all networks when there is any dangerous earth leakage; this in turn can result in a downtime excavator [35].

Consequently, for the safe and efficient operation of the relay leakage, AIMD must review the principles of power supply with users of the excavator and lead parameters of electricity networks in compliance with the technical data relay.

Measurements on the coal mine Ekibastuz were taken by the developed methodic of determining the insulation parameters in networks with isolated neutral under 1000 V at normal operating conditions of the electrical network of excavators with operating electrical receivers [8].

The developed methodic is based on the method of determining the parameters in the short network with isolated neutral. The method consists of measuring the modulus of linear voltage, phase voltage in respect to earth after the connection between phase and earth the auxiliary conductance. From the measured values of the modulus of linear stress, voltage to earth after the connection between phase and earth the auxiliary conductance, bear the auxiliary conductance in mind, admittance, conductance, and capacitive susceptance of phase-to-ground are determined with satisfactory accuracy [6, 36].

According to the measured values of the modulus of linear voltage, U_l, phase voltage in respect to earth, U_{pho}, when connecting auxiliary conductance, g_o, is determined admittance, conductance and capacitive susceptance of isolation by mathematical dependences [37]:

- admittance of isolation

$$y = \frac{1.73U_{pho}}{U_l - 1.73U_{pho}} g_o, \tag{24}$$

- conductance of isolation

$$g = \left(\frac{3U_{\text{pho}}^2}{U_l^2} - \frac{3U_{\text{pho}}^2}{\left(U_l - 1.73U_{pho}\right)^2} - 1 \right) 0.5 g_o, \tag{25}$$

- capacitive susceptance of isolation

$$b = \left(y^2 - g^2\right)^{0.5}. \tag{26}$$

Based on the results of the determination of admittance, conductance, and capacitive susceptance of isolation in a short line 0.4 kV on the excavator EKG-8I in the coal mine Ekibastuz, Angrensor LLP, the results were processed by using the small sample method. The results from the experimental research regarding the parameters of the insulation as well as an assessment of the results collected using the small sample method are shown in **Table 1**.

Studying the isolation of electric networks under 1000 V of the excavator showed that the insulation resistance is due to active resistance which characterizes the properties of the dielectric of insulating material used for insulation of live parts

Parameters of the insulation	The number of measurements								X mean value of parameters
	1	2	3	4	5	6	7	8	
Admittance of isolation, $y \times 10^{-5}$, Ohm	2.20	2.21	2.18	2.24	2.17	2.22	2.15	**2.17**	2.19
Conductance of isolation, $b \times 10^{-5}$, Ohm	1.34	1.35	1.37	1.40	1.39	1.41	1.35	**1.37**	1.37
Capacitive susceptance of isolation, $g \times 10^{-5}$, Ohm	1.74	1.75	1.69	1.75	1.67	1.71	1.67	1.68	1.71

Table 1.
The results of determination of the parameters of the insulation in a short line 0.4 kV on excavator EKG-8I in the coal mine Ekibastuz, Angrensor LLP.

Z, Ohm	R, Ohm	X, Ohm	$\tan\delta$, –	I_o, mA
$\frac{45662}{44643–46512}$	$\frac{58480}{57143–59880}$	$\frac{72993}{70922–74627}$	$\frac{0.80}{0.80–0.81}$	$\frac{4.93}{4.84–5.04}$

Table 2.
Numeric values of parameters of the insulation and the current of single phase-to-earth fault in the network under 1000 V on the excavator EKG-8I. Z – impedance, R – resistance, X – reactance, tanδ – loss tangent of a dielectric, I_o – single-phase earth current.

of the conductors with respect to the ground. Capacitive resistance is higher than active resistance of insulation in networks under 1000 V. As such, the current of single phase-to-earth fault under 1000 V on an excavator is not due to a capacitive component but an active component. Experimental studies showed that the current of a single phase-to-earth fault in the network under 1000 V on the excavator EKG-8I of coal mine Ekibastuz, Angrensor LLP, is about 5 mA. The current of the single phase-to-earth fault in the network under 1000 V on the excavator has a lower value than the set point of RCD. In light of this, it is obvious that the RCD used on excavators by their specification does not provide people with effective protection from electric shocks in networks under 1000 V.

The determined absolute value of the expected capacity of the network is 4.30×10^{-5} uF, and the range of the network capacity under 1000 V is $4.20 - 4.42 \times 10^{-5}$ uF based on the data in **Table 2**.

According to the received data, the parameters of the insulation electrical short network are changed insignificantly and are at a high level. This can be explained by the fact that the power supply circuit of the excavator does not contain a network with distributed parameters and the capacity of the network consists of a phase-to-earth capacitance only for the electrical receiver. This stipulates for high loss-angle tangent of isolation in the network under 1000 V on the excavator EKG-8I. The current of the single phase-to-earth fault in the network under 1000 V on the excavator EKG-8I has a small value and a small range of variation. There is a range of variation when it comes to the parameters of network isolation, loss-angle tangent of isolation, and current of single phase-to-earth fault in the network under 1000 V due to changes in the supply voltage [38].

Based on the aforementioned evidence, it follows that the workers of mining enterprises receive electric shocks during excavator maintenance work due to the ineffectiveness of RCD. Indeed, as the RCDs used on excavators do not work, this leads to a violation of safety rules regarding the use of electrical equipment of excavators in mining companies. As such, it is necessary to develop RCD for three-phase mains under 1000 V on excavators and develop technical measures to increase efficiency of RCD on excavators.

4.3 The method to improve efficiency of residual current device

The development of RCD for a three-phase electric network under 1000 V on excavators is a complex and expensive task, as the principle of operation of RCD for the three-phase electric network with isolated neutral under 1000 V must be changed.

RCDs such as AIMD are used on excavators of mining enterprises. The principle of their operation was based on Scheme 3B developed by Professor. Leybov in the 1950s. This principle is still used today and was thoroughly studied by Shishkin in the 1960s at the Skochinsky Mining Institute. All these studies were conducted to improve the efficiency of RCD in mining electric networks under 1000 V. With this said, however, no detailed investigation into the electric network under 1000 V on excavators has been carried out [39]. In light of this, the most vital area relates to the development of technical measures with which to improve the effectiveness of RCD on excavators, taking into account the study of insulation parameters in short electric networks under 1000 V.

There is a drawback when it comes to the existing method of RCD in the network with isolated neutral under 1000 V on excavators. Indeed, this method is based on setting-up a direct current into three-phase mains with a fixed set-point of protection from electric shock. The disadvantage is that the fixed set-point of current of protection does not protect people from electric shocks, as short networks under 1000 V on the excavator have a current of single-phase earth fault, which is less than the set-up value of RCD. In order to overcome this problem, it is necessary to develop a method to improve the efficiency of RCD in a network with isolated neutral under 1000 V on excavators.

Improving the efficiency of RCDs in a network with isolated neutral under 1000 V is based on switching off the supply due to increase in the phase capacity with respect to earth when insulation is damaged.

A method for improving the effectiveness of RCD in a network with isolated neutral under 1000 V on excavators is explained by the electrical circuit diagram found in **Figure 11**. The circuit diagram contains: a power transformer T; load interrupt switch QF1, which supplies voltage to the three-phase electric network; three-phase electric network with the phases A, B, and C; electrical receivers; load interrupt switch QF2, which switches capacitors between phases of network and ground; capacitors C1, C2, and C3, being provided by an increase in the current of single-phase ground fault; load interrupt switch QF3, which switches residual current device; residual current device—RCD; total admittance of network isolation Z1, Z2, and Z3 [40].

The principle of operation of the scheme of safety shutdown in the short electric network with insulated neutral under 1000 V on excavators is as follows: the power is supplied to three-phase electric network with the phases A, B, and C from power transformer T by load interrupt switch QF1, where electrical receivers are supplied with voltage under 1000 V of excavator. The capacitors C1, C2, and C3 are connected by load interrupt switch QF2 between the electric network phase and earth to provide increased current of single-phase earth fault. The RCD is connected to a three-phase electric short excavator network by load interrupt switch QF3 [41].

An RCD with a fixed set-point does not allow for the shutting off of the three-phase electric network by load switch QF1 when any phase-to-earth insulation of network is damaged. Thus, there is the risk of electric shock. The RCD does not turn off the three-phase electric network when any phase-to-earth insulation is damaged, as the set-up point of current for protection is more than the current of

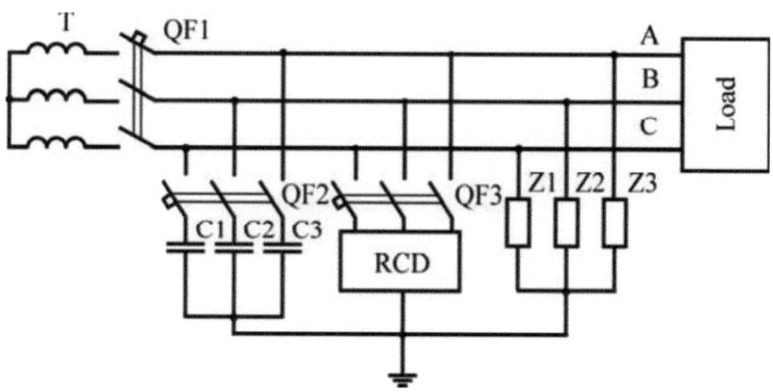

Figure 11.
The scheme of residual current device in electric network with isolated neutral under 1000 V on excavators.

single-phase earth faults in the three-phase network of the excavator. In order to disable the three-phase network when insulation is damaged, the current of single-phase fault in the network is increased by means of connecting capacitors C1, C2, and C3 between the phases of the electric supply and the ground by load switch QF2. In this case, the current of single-phase circuits in the excavator's three-phase network will be more than the current of the set-up point of RCD, which will activate the RCD. Thus, the switching off is made possible thanks to the load interrupter switch QF1 supplying voltage from the power transformer [40].

Implementation of the developed method to improve the effectiveness of RCDs in electric networks under 1000 V will ensure the growth of level of electrical safety when using electrical installations and reduce the number of accidents on excavators.

5. Conclusion

The following results were obtained in this work:

1. A method for determining the parameters in three-phase networks with isolated neutral voltage up to 1000 V and above is to measure the modulus of the line voltage and phase voltage with respect to ground and A, and after you connect, an additional active conductivity between the A-phase mains and earth was developed.

2. Error analysis of method for determining the parameters of isolation in three-phase electrical network with isolated neutral showed that it is necessary to select a certain value of additional active conductance, so as to ensure satisfactory accuracy required when determining the:

    - the total conductance of mains phase insulation relative to the ground is chosen such additional active conductance, the values were within = 0.2–0.8, with the error does not exceed 5% when using measuring devices with accuracy class 1.0, and 2.5% using measuring devices with accuracy class 0.5;

    - active conductance in three-phase electrical network with isolated neutral voltages up to and above 1000 V select such active additional conductance g_o, so that U_{A*} = 0.2–0.8, when U_{C*} = 0.2–0.8, then the error does not exceed 3.5% when using the measuring devices with accuracy class 1.0;

    - capacitive conductance of electrical network phase insulation relative to the ground select such additional active conductance g_o, so that U_* = 0.2–0.8, with the change tan δ = 0.6–1.6, then the error does not exceed 5% when using the measuring devices with accuracy class 1.0, and 2.5% when using the measuring devices with accuracy class 0.5.

3. The developed methods provide satisfactory accuracy, simplicity, and security in its implementation in the three-phase electrical networks with isolated neutral voltages up to and above 1000 V.

4. The chapter presents new evidence-based results that solve the important scientific task of ensuring electrical safety in networks with an isolated neutral

voltage up to 1000 V in mining enterprises through the development of methods to control the condition of insulation.

A method of measuring the admittance of network with an isolated neutral voltage up to 1000 V is based on the measurement of the modulus of the zero phase-sequence voltage and phase voltage to earth, with an additional conductance where the value of the regulation is made additional conductance in conduction to ensure the equality of the modulus of phase voltage to earth and zero phase-sequence voltage. In ensuring the equality of zero phase-sequence voltage and phase voltage to earth connection of additional conductance, it corresponds to the admittance network isolation.

The simulation model of method of measuring the admittance of insulation in the Matlab/Simulink environment was modulated. The developed model allows for the regulation variable resistor to be used to simplify the calculations of the parameters of network isolation. Due to the data received in the regulation of the variable resistor to 2068 Ohms, the true value zero phase-sequence voltage and phase A voltage to earth are equal to 140.8 V. Thus, the variable of the admittance y corresponds to 2068 Ohms, which comprises 0.48 mS.

Developing a method of measuring the admittance of insulation networks with an isolated neutral voltage up to 1000 V will provide improved accuracy and speed measurement admittance network isolation. The proposed method is simple, as the instrumentation, single-phase voltage transformers, required for measuring the admittance network isolation is in the service manual enterprise energy management.

5. The experimental data obtained are composed of numerical values of the parameters of the insulation on the excavator EKG-8I of coal mine Ekibastuz, Angrensor LLP. It was established that the insulation resistance is due to active resistance, which characterizes the properties of the dielectric of insulating material used for insulation of live parts of the conductors with respect to ground. Capacitive resistance is higher than active resistance of insulation in networks under 1000 V.

It is found that the RCDs used on excavators by their specifications do not provide effective protection from electric shocks in a short network with voltages under 1000 V as the current of single-phase earth faults in the network under 1000 V on the excavator has less value than the current of the RCD set-up point.

A new method aimed at improving the effectiveness of RCDs in electric network under 1000 V has been developed and is based on setting up the DC into a three-phase network with a fixed set-point of protection from any phase-to-earth insulation damage, where the equipment is switched off by residual current device when live-line bare-hand touching of electric equipment occurs. This is due to increases in the phase capacity with respect to earth.

Organizational and technical measures aimed at improving the reliability and level of electrical safety in electrical mining enterprises will help to protect people from electric shocks while also reducing the number of accidents at work.

The work was carried out in accordance with the contract no. 242 of March 17, 2018, at the S. Seifullin Kazakh Agro-Technical University with the Ministry of Education and Science of the Republic of Kazakhstan under the project no.

AP05132692 "Development of innovative technologies for increasing the efficiency of power supply for electric receivers with voltages up to 1000 V at mining enterprises."

Author details

Bolatbek Utegulov
Energy Department, S. Seifullin Kazakh Agro-Technical University, Astana, Republic of Kazakhstan

*Address all correspondence to: utegulov.b@mail.ru

References

[1] Gladilin LV, Shutski VI, Batsezhev Yu G, Chebotayev NI. Elektrobezopasnost' v gornodobyvayushchei promyshlennosti [Electrical Safety in the Mining Industry]. Moscow: Nedra Press; 1977. p. 327

[2] Utegulov BB, Zhumadirova A. In: Utegulov BB, editor. Microprocessor Means for Monitoring of Insulation Condition and Protection from a Single-phase Earth Fault in Electrical Networks of 6–10 kV. Pavlodar: S.Toraighyrov Pavlodar State University; 2005. p. 151

[3] Bapeshev Yu G, Utegulov BB. Three-phase Network Single-phase Earthing Determn by Determn Phase Shift between Additional Capacitive and Zero Sequence Currents. Patent Number: SU1626218-A; Patent Assignee: RUDNENSK IND INST

[4] Utegulov BB. Neutral Three-phase Single-phase Ground Fault Protector has Matching Transformer Primary Connected via Resistor to Zero Sequence Current Transformer. Patent Number: SU1403189-A; Patent Assignee: Utegulov BB

[5] Utegulov BB. Isolated Neutral Three-phase Circuit Insulation Parameter Determn Using Known Conductance between Phase Circuit and Earth and Measuring Phase Shift between Line and Neutral Sequence Voltages. Patent Number: SU1347038-A; Patent Assignee: RUDNENSK IND INST

[6] Shutski VI, Utegulov BB. Three-phase Insulated Neutral Network Insulation Testing by Measuring Zero Sequence and Phase Voltages of Network with Additional Conductance in One Phase. Patent Number: SU917127-A1; SU917127-B; Patent Assignee: PAVLODAR IND INST; MOSC MINING INST

[7] Utegulov BB, Utegulov AB, Begentaev MM, Begentaev BM, Uakhitova AB, Zhumazhanov S, et al. Method for determining parameters of isolation network voltage up to 1000 V. In: Mining Enterprises Source of the Document Proc. of the IASTED Int. Conf. on Power and Energy Systems and Applications (PESA); Pittsburgh, USA. 2011

[8] Utegulov BB, Utegulov AB, Begentaev MM, Begentaev BM, Uakhitova AB, Zhakipov N, et al. Method for determining the insulation in asymmetric networks with voltage up to 1000 V. In: Mines Proc. IASTED International Conference on Power and Energy Systems and Applications (PESA); Pittsburgh, USA. 2000

[9] Sinclair J. Quarrying Opencast and AlluvialMining. ISBN 978-94-011-7611-8. Netherlands: Springer; 1969

[10] Mattis AR, Cheskidov VI, Labutin VN. Choice of the hard rock surface mining machinery in Russia. Journal Mining Science. 2012;**48**(2):329-338

[11] Fry PR. Loads and stresses—The real cause of failures in surface mining machinery. Practical Failure Analysis. 2003;**3**(2):7-14

[12] Xiangjun Z, Li KK, Chan WL, Xianggen Y. On-site safety evaluation for earth fault in mining power systems. Browse Journals & Magazines. 2003; **39**(6):1563-1569

[13] Pravila tekhniki bezopasnosti dlya elektricheskih ustanovok [Safety Regulations for Electrical Installations]. Moscow: Glavgosenergonadzor Rossii; 2011

[14] Pravila bezopasnosti v ugol'nykh i slantsevykh shakhtakh [Safety Rules for Coal and Shale Mines]. Moscow: Nedra; 1986

[15] Dunkle WF. Insulation field test results. Browse Journals & Magazines. 1951;**70**(7);579-579

[16] Instruktsiya po bezopasnoi ekspluatatsii elektrooborudovaniya i elektrosetei na kar'erakh [Instructions for the Safe Operation of Electrical Equipment and Power Grids in the Quarries]. Moscow: Nedra; 1982

[17] Utegulov BB, Utegulov AB, Uakhitova AB. Development of method to improve efficiency of residual current device under 1000 V on excavators of mining enterprises. Journal of Mining Science. 2016;**52**(2):325-331

[18] Pravila ustroistva elektroustanovok [Electric Installation Code]. Novosibirsk: Novosibirsk State Technical University; 2007

[19] Pravila tekhnicheskoi ekspluatatsii ugol'nykh i slantsevykh shakht [The Technical Operation of Coal and Shale Mines]. Moscow: Nedra; 1976

[20] Abalakov GI. Research and development of pre-insulation monitoring for high voltage cables in mines [Cand. Tech. Sci. dissertation]. Kemerovo; 1999

[21] Sidorov AI, Utegulov BB, Utegulov AB, Uakhitova AB. Procedure for Isolation Parameters under Damage of a Phase in a Mine Network up to 1000 V (Tr. KarGTU, no. 3). 2007. pp. 88-90

[22] Yagudaev BM, Shishkin NF, Nazarov V. Zashchita ot elektroporazheniya v gornoi promyshlinnosti [Protection Against Electrocution in Mining Industry]. Moscow: Nedra; 1982

[23] Valinevicius A, Keras E, Balaisis P. Protection against electric shock using residual current devices in circuits with electronic equipment. Electronics & Electrical Engineering. 2007;**76**:51-54

[24] Xiang Y, Cuk V, Cobben J. Impact of residual harmonic current on operation of residual current devices. In: Proceedings of 10th International Conference on Environment and Electrical Engineering (EEEIC); Rome. 2011

[25] Meijer C, van der Ven J-K, Ross R. EMC and electrical safety on board ships: How EMI filters undermine the protection against electric shock. In: Proc. Int. Symp. on Electromagnetic Compatibility (EMC Europe) (Brugge); 2013

[26] Kisarimov RA. Elektrobezopasnost [Electrical Safety]. Moscow: Radiosoft Press; 2013

[27] Manoilov VE. Elektrobezopasnost' i chelovek [Electrical Safety and Man]. Moscow: Nauka; 1991

[28] Gladilin LV, Utegulov BB. Accuracy analysis of method for determining insulation parameters in a three-phase electrical networks with isolated neutral by voltage above 1000 V. Proceedings of the Higher Educational Institutions. Mining Journal. 1980;**8**:94-97

[29] Zaydel' AN. Elementary Estimates of Measurement Errors (Revised and Updated). Vol. 3. Leningrad: Nauka; 1968. p. 97

[30] Nagornukh VV. Matematicheskoe modelirovanie opredeleniya parametrov izolyatsii otdel'nykh faz elektricheskoi seti s izolirovannoi neitral'yu [Mathematical Modeling of Parameters of Isolation of Individual Phases in Electrical Networks with Isolated Neutral] (GIAB: no. 12). 2011. pp. 394-405

[31] Utegulov AB. Razrabotka fazochuvstvitel'nykh metodov povysheniya urovnya elektrobezopasnosti i nadezhnosti ekspluatatsii elektricheskikh setei s izolirovannoi neytralyu napryazheniem

do i vyshe 1000 V [Development of phase-sensitive methods to enhance safety and reliability of electrical networks with isolated neutral voltage up to 1000 V and above] [Cand. Tech. Sci. dissertation]. Pavlodar: PSU; 2003

[32] Utegulov BB, Utegulov AB, Uakhitova AB, Begentayev BM, Amurgalinov SA. Sposob izmereniya polnoy provodimosti izolyatsii v seti s izolirovannoy neytral'yu napryazheniyem do 1000V [Method for Measuring the Total Conductivity of Insulation in a Network with an Isolated Neutral Voltage of up to 1000 V]. Innovation Patent No. 24286. Astana: KazPATENT; 2011

[33] SimPowerSystems User's Guide COPYRIGHT 1998–2003. TransÉnergie Technologies Inc., under sublicense from Hydro-Québec, and The MathWorks, Inc

[34] Chernykh IV. Modelirovanie elektrotekhnicheskih ustroistv v MATLAB [Simulation of Electrical Devices in MATLAB). SimPowerSystems and Simulink]. Moscow: DMK Press; 2008

[35] Utegulov BB, Utegulov AB, Uakhitova AB, Amurgalinov ST, Begentaev BM. Issledovanie bezopasnosti proizvodstva rabot v setyakh napryazheniem do 1000 V na ekskavatore EKG-8I [A Safety Study of Works in Networks up to 1000 V on excavator EKG-8I] (Vestn. PSU. Series on Energy: no. 3); 2009. pp. 111-114

[36] Uakhitova AB. Isolation Control Procedure for Asymmetrical Network with Isolated Neutral and Voltage of 6–20 kV (Izv. Vuzov, Probl. Energetik: KGEU Nos. 9–10). 2012. pp. 102-106

[37] Utegulov BB, Utegulov AB, Uakhitova AB, Amurgalinov ST, Begentaev BM. Numerical Values of the Insulation of Electrical Networks up to 1000 V Excavator EKG-8I (Vestn. PSU. Series on Energy: no. 4); 2009. pp. 102-107

[38] Edinye pravila bezopasnosti pri razrabotke mestorozhdenii poleznykh iskopaemykh otkrytym sposobom [Common Safety Regulations for Open Pit Mineral Mining]. Moscow: Gostekhnadzor; 2003

[39] Lazarev AI. Development of safety shutdown system for mine electric networks under 1 kV with a frequency-controlled electric drive [Cand. Tech. Sci. dissertation]. Moscow: MEI; 1998

[40] Utegulov BB, Uakhitova AB, Utegulov AB, Amurgalinov ST. Sposob zashchitnogo otklyucheniya v seti s izolirovannoy neytral'yu napryazheniyem do 1000 V na ekskavatorakh [Method of Protective Shutdown in a Network with an Isolated Neutral Voltage of Up to 1000 V on Excavators]. Innovation patent no. 23240. Astana: KazPATENT; 2011

[41] Utegulov BB, Uakhitova AB, Utegulov AB, Amurgalinov ST. Method of Safety Shutdown in Electric Network with Isolated Neutral under 1000 V on Excavators. Vol. 1. Kazakhstan: Nauka Tekhn; 2010. pp. 105-107

Chapter 5

An Integrated Approach for the Building and the Selection of Multidisciplinary Teams in Health Care System

Ikram Khatrouch, Lyes Kermad, Anderrahman El Mhamedi and Younes Boujelbene

Abstract

This chapter presents a support approach for the building and the selection of multidisciplinary teams. In the first part, we propose a new general model for multidisciplinary team building. The proposed model takes professional's preferences into account when a team building process is required for any type of project. In the second part, we develop hybridization between a multicriterion decision method and a cognitive method for selection teams. The developed methodology is based on the experiences of the past operations in order to select the adequate team for a new operation. We test the effectiveness of the model using health care domains of different complexities and describe some practical experiences of using the model in the surgical team building process.

Keywords: team building, multidisciplinary team, selection team, analytic hierarchy process, case-based reasoning

1. Introduction

Hospital context has undergone multiple changes in the last decade economical field (expenditure growth of hospital logistics), technological field (integration of new technologies), and social field. In this context, hospital systems aimed to reduce expenditure while ensuring greater quality of care. Adverse events related to care affect 3.7–16.6% of patients care in the OECD countries. The incidence in France is 5.1% which represents an average of 6.6% of the adverse events for 1000 hospital days [1]. Also, around half of events occurs during surgical intervention [2, 3], which represents the emblematic of this component.

Investigations focus on complications in the operating rooms date from the 1980s. It is thus crucial to understand causes of complications. Several studies draw our attention for research on the causes of surgical complications. Atul et al. [4] have shown in their study about three hospitals that two-third of complications produced during operative phase. Three factors were cited as factors that contribute to error: the lack of experience/lack of competence for

Studies	The causes				
	Communication	Leadership	Technical skills	Fatigue/excessive workload	Collaboration
Atul et al. [4]	*		*	*	
Helmreich and Schafer (1994)	*				
Watson et al. [8]	*		*		
Taylor et al. [9]			*	*	*
JCSES [6]	*				
Wong et al. (2009)	*	*			
Fuchshuber et al. [7]	*				
Haller et al. (2011)	*				
Doppia et al. (2011)				*	

Table 1.
The causes of complication in the operating room.

surgical task was associated to 53% of incidents, communication problem (43%), fatigue or excessive working (33%). Fleming et al. [5] analyzed interpersonal skills for each members of cardiac surgery team to determine their attitudes regarding team work. Researchers have identified different factors as seniority in grade and adherence in professional community. Respondents reported that erroneous communication, execution of intervention at the wrong time and the not following of procedures that constitute the most frequent types of errors. Statistics of JCAHO show that 65% of severe events (for example, compresses forgotten, error in blood transfusion, etc.) are related to a lack of communication [6].

The results of the analysis of National Surgical Quality Improvement Program (NSQIP) identified the major problems like communication team, lack of skilled care during the patient postoperative care. It is important to note that problems are related to systems and not to individual performance of surgeon [7].

We often classify in the table below different studies depending on the causes identified.

Improvement of techniques and processes in the operating theater does not completely solve complications occurring. Refer to the studies mentioned (**Table 1**), we can conclude that team building constitutes an obvious starting point.

2. Domain and motivation

Operating theater is a containment with high concentration of human competence. An operation needs intervention of different actors from different disciplines (surgeons, anesthetists, nurses, etc.) with various levels of skills. Surgical team performance emerges as key points to ensure the best quality care and risk management. The operating theater is also a deep human place where the individual works on an individual and with an individual. These individuals have personalities, logic, interests, and specific different viewpoints and sometimes conflicting. They constitute a surgical team in which performance and outcomes depend on the degree of coordination the efforts made by everyone, that is, teamwork. Selection teams ensure that the right team is in place and that it will have a capable leader in place.

Successful building and selection teams are still an open problem in various fields of social, business, and hospital studies. To solve this problem, several methods were proposed such as AHP [10, 11], fuzzy-genetic algorithm [12], multiobjective optimization [13], fuzzy logic (Shipley et al., 2013), etc.

The main objective of this chapter is to propose a systematic evaluation model to help the decision maker for the building and selection of an optimal team among a set of available alternatives. For building team, we present a new algorithm applied to multidisciplinary team. Then, for selection team, we envelop a methodology where we combine a multicriterion decision method and a cognitive method.

The remainder of this chapter is organized as follows: In Section 3, proposed model for weapon building and selection team is presented and the stages of the proposed approach are explained in detail. In Section 4, experimental results and data analysis are discussed. Finally, conclusions of this study are made in Section 5.

3. The model description

Proposed model is divided in two main parts: approach support for building multidisciplinary teams and approach for selection teams, presented in **Figure 1**.

These two approaches can be applied successively or separately, depending on the case of application.

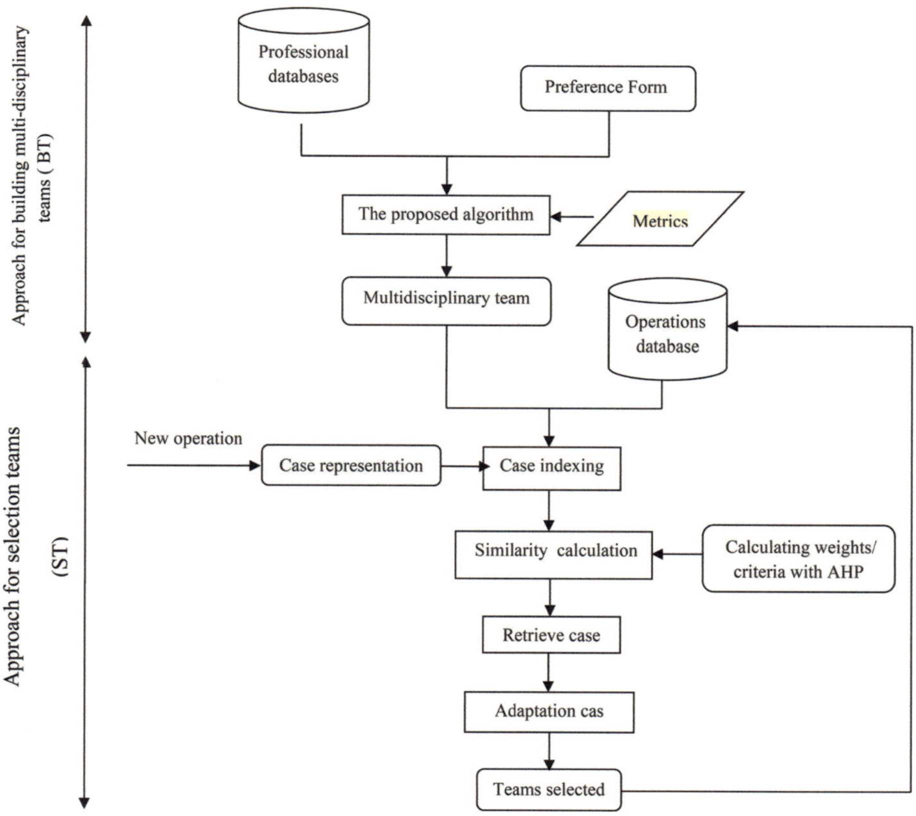

Figure 1.
Model support for building and selection teams.

3.1 Approach support for building multidisciplinary teams

This first approach presents a new model for multidisciplinary team building. It takes professional's preferences into account when a team building processes. The proposed approach is presented in four main steps explained below.

Step 1: Completion of preference form

At the beginning of the year, the professionals in the operating theater (proposers) are asked to complete a form (**Figure 2**) for ranking their colleagues (acceptors) using a preference scale from 1 to 6 (1 being highest and 6 being lowest) according to their willingness to be in the same group. This process should be finalized within a period of 7 days. Although proposers are completing the forms, they should agree to the following rules:

1. All professionals must submit a form at the beginning of the year; otherwise the proposer agrees that all the acceptors will be regarded as having the same priority with the highest level.

2. They cannot give the same preference order for more than one acceptor.

Step 2: Constructing the preference matrix

We transform the forms into a preference matrix. Several revisions are made on the matrix according to Assumptions 1 and 2.

Name: Id: Discipline:					
Surgeons	Order preference	Anaesthesists	Order preference	Instrumentalists	Order preference
S1		A1		I1	
S2		A2		I2	
Si		Ai		Ii	
Sn		An		In	

Preference Form

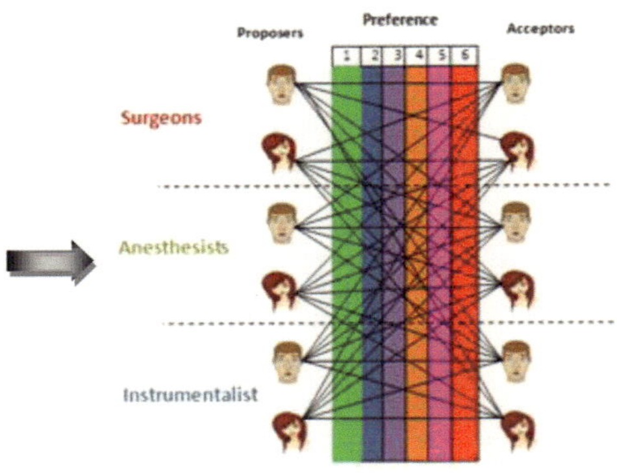

Preference matching

Acceptor/ Proposer		Surgeons			Anaesthesists			Instrumentalists		
		S1	Si	Sn	A1	Ai	An	I1	Ii	In
Surgeons	S1									
	Si									
	Sn									
Anaesthesists	A1									
	Ai									
	An									
Instrumentalists	I1									
	Ii									
	In									

Preference Matrix

Figure 2.
Transfer sequence of the preference forms to a preference matrix.

Assumption 1. If a proposer professional does not complete a form, the proposer would accept being a member of any team without complaint. Thus, the rows of these proposers in the matrix are filled with 1 (highest preference level) for each acceptor (column), P(j) = 1, for j = 1 to n−1.

Assumption 2. If an author does not give a priority level for acceptors, the author agrees that all these acceptors have the same priority level. Thus, the priority level of the acceptor (columns) is set to the lowest priority given by the author plus 1. P(j) = P(i) + 1 for j = i + 1 to n−1.

When the preference matrix is constructed, it is transformed into a lower triangular matrix by adding the weights of each cross proposer and acceptor ($M_{ij} = M_{ij} + M_{ji}$).

Step 3: The team building algorithm

The algorithm of the proposed model is straightforward, and it is similar to Prim's minimum spanning tree algorithm and Sahin algorithm. **Figure 3** presents the pseudocode of the algorithm. We begin by traversing all the elements of the first discipline in order to find the two groups who have the minimum weights. For this, we use the function FindMinRelationShip. The function chekGroupComposition permits to verify if it is possible to merge the two groups (e.g., if we merge the two groups, the total number of surgeons is less than maximum surgeon authorized in one group). If the merging of the groups is possible, we remove the second group and we recalculate the new weights. If the merging is not possible, we put a negative value in a matrix of preferences. We repeat the same steps until all the weight values are negative. When this first phase is finished, i.e., we can no longer create a new group using the first discipline, we add the individuals of the second discipline. We recall the same function, FindMinRelationShip, chekGroupComposition, merge, until no way to merge groups. This last phase is repeated until all disciplines are added.

A sample example is presented in **Figure 4**, for application of the proposed algorithm. Suppose we have nine employees with three disciplines (three surgeons, three anesthetists, three instrumentalists) and we need to compose teams with three members (one surgeon, one anesthetist, and one instrumentalist). We apply the algorithm above; we obtain this composition (Step 4—**Figure 4**) of three teams.

This developed approach represents an improvement for Sahin algorithm [14]. We have developed our computer algorithm on the Java platform within the Eclipse. The next step of procedure is selection teams, as detailed below.

3.2 Approach for selection teams

Once we have teams already built, we are going to apply this second approach that helps the decision maker to find more appropriate team; which means, the team that is adapted to his preferences and the need of each operation. The proposed model is presented in four main steps explained below.

Step 1: Case base construction

The presentation of the base depends strongly on the structure and content of such cases. A case base contains problems and solutions that can be used to derive solution for a new situation. In our work, cases contain a vector of attributes that define the problem and the solution, which correspond to the best team that satisfies exactly the needs of the operation and the preferences of the decider. A case is described by the criteria and also the solution.

```
Algorithm 1: Multidisciplinary Team Building algorithm

Input: Number_discipline,              //number of discipline
 Number_indiv_by_type[],               // number of individual by discipline
Max_indiv_by_Type[],                   // Max of individual in team for each discipline
Nb_total_individu;                     // Total number of individuals
Preference Matrix Mij,
Output: groups
   For j : 1.. NbGroup
        Maxpref= Max(M[j]) ;
        For  i :1..NbGroupe do
          If M[i,j] = 0 then Set M[i,j] = maxpref+1
         end {for}
  end {for}
Transform (M)                          // each individual is considered as team composed by a single person
For ( i=0, i< Number_discipline , i++)   //We construct the team by using the first discipline
  For (j =0,j<nbgroupe,j++)            //  by using the same algorithm used by sahin to find all possible team.
         FindMinRelationShip (group1, group2, weight)  //When we finish with first discipline
                    //we add the all group of the second discipline and so forth until the last discipline

        IF checkGroupComposition (group1, group2,Max_indiv_by_Type[])
                Merge (group1, group2)
                Remove (group2)
                NbGroupe - -
        Else M[g1,g2] = -1
    end {if}
  B=False
For i : 1..NbGroup
    For j : 1.. NbGroup
          If (M[i, j] <> -1) B=True;
           break;
    end {For}
  If (B==True) Break;
end {For}


         If (B==True) Break;
       end {For}
     If (B==False) Break;
     end {For}
     end {For}
     Chek GroupeComposition (groupe1, groupe2, Max_indiv_by_Type[])
     For ( i=0, i< Number_discipline , i++)      //we check for each type if the sum the individuals in

         If (groupe1.nbre(i)+ groupe2.nbre(i)) >Max_indiv_by_Type[i])  //Groupe 1 et Groupe 2 is
             return false;     //greater than the maximal number authorized for this discipline we return False
         end{IF}
     end {For}
             return true;
```

Figure 3.
The algorithm's pseudocode.

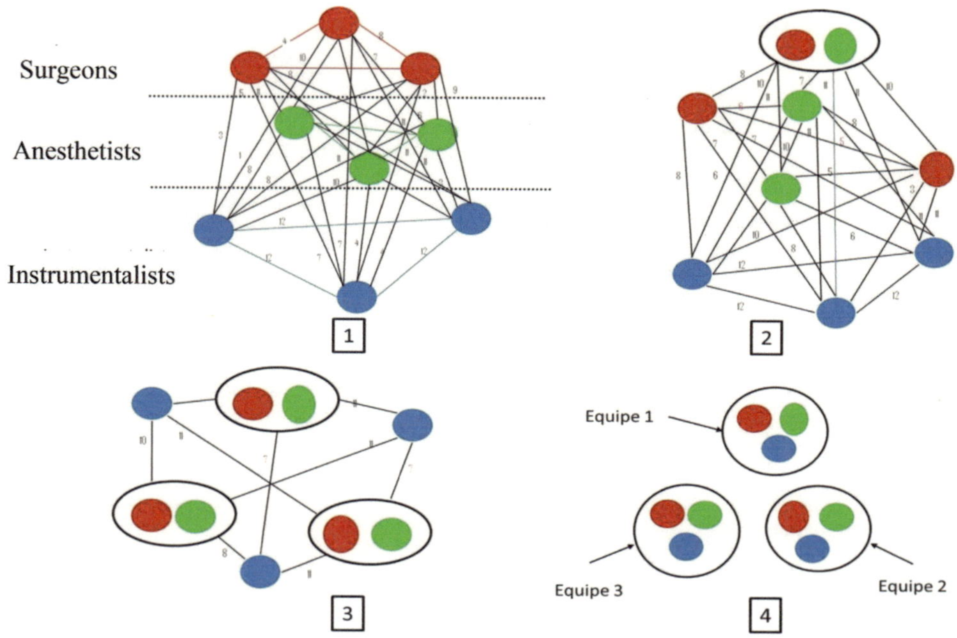

Figure 4.
Sample steps for the multidisciplinary team building algorithm.

Criteria:

The criteria which characterize the team choice are:

- The time (T): the duration of the operation
- The competence (Ct): the technical competence of the team.
- The communication (Co): the communication in the team.
- The risk criticality (R): the criticality degree of the risk.

Solution:

It is represented by the best team which satisfies exactly the needs and the preferences of the decision maker. That is defined by a set of criteria.

Step 2: Calculate the weights of criteria

In this step, the AHP method is used to determine the weights of criteria for case similarity analysis. This weight is the key to case retrieval. For this reason, we use the analytic hierarchy process (AHP) to determine the relative weight of each attribute according to its importance and use these important weights to calculate the similarity among the new coming case and each case in the case base.

The first step is to compose our problem in three hierarchical levels presented by **Figure 5**.

The next step is to conduct a questionnaire survey handed to each member. The value assigned is based on the scale in interval of 1–9. Then, create square pair-wise comparison matrices of the selection criteria. **Table 2** [15] presents the scale of preference in the pair-wise comparison process.

The consistency of results obtained is found by calculating the consistency index (CI). More consistency index becomes bigger and more the judgments of the user are coherent and vice versa.

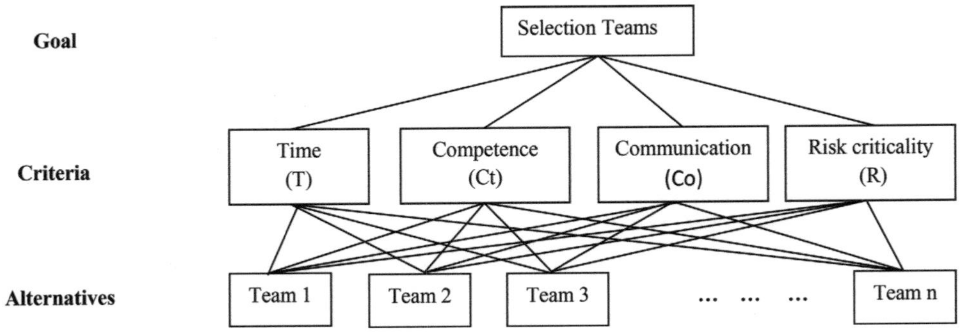

Figure 5.
An AHP structure for selection teams.

Verbal judgments	Numerical rating
Equal importance	1
Moderate importance of one over another	3
Strong or essential importance of one over another	5
Very strong importance	7
Absolute importance	9
Intermediate values between two adjacent judgments	2, 4, 5, 8

Table 2.
AHP comparison scale.

Step 3: Retrieving phase

The objective of the retrieving phase is to find the most similar previous cases in case base, and retrieving them for analysis, in order to select one and reuse it in the next phase. The similar case retrieval depends on the case representation and their indexing in the case base. The objective is to measure the similarity between the new case (operation) and the stored cases in the case base.

The question in our model is which one of the previous teams is the most similar to the new operation (case) that must be treated. In order to evaluate the similarity, the similar attribute collection $S = \{sT_1,..., sT_n\}$ should be determined first. Let us denote the new operation (case) to be considered by T′. By T, we denote operation (case) stored in the case base. We also denote by Sim the similarity degree between the new operation and the operation stored.

In the first step, we calculate the local similarity sT_i between attribute. We define this similarity in the following way:

$$sT_i = \left(1 - \frac{\left|T_i - T_i^{'}\right|}{T_i^{max} - T_i^{min}}\right) \quad (1)$$

where T_i is the ith attribute of the case in memory, T'_i is the ith attribute of the current case, and T_i^{max} and T_i^{min} are the maximum and minimum values between all the cases for the ith attribute.

In the second step, we calculate overall similarity by using the weights associated with each attribute. We thus introduce the importance of the attributes as a new variable. It measures the importance of the ith attribute, which we express as T_i. In our model, the weights W_i were calculated by using the AHP method. A general form of similarity measure function is shown in Eq. (2).

$$\mathrm{Sim}(\mathrm{T},\mathrm{T}') = \frac{\sum_{i=1}^{n} sT_i * W_i}{\sum_{i=1}^{n} W_i} \quad (2)$$

where T is the case in memory, T′ is the target case, and n is the number of attributes of each case. Finally, the case having the biggest global similarity with the new case will be selected.

Step 4: Construction of the new case solution

The objective of this phase is to evaluate the retrieved solution. Thus, the decision maker must judge if the selected case is well or no. If yes, this case solution will be adapted to the new case. Otherwise, he passes to the second more similar case, to the third, etc. Finally, the new case and its validated solution are integrated into the case base. It is then necessary to know which information can be important to retain, how to index the case for a future retrieve, and how to integrate the new case in the case base.

4. Experimental results and data analysis

To assess the computational tractability and efficiency of the developed model, we tested the operation of our model on a set of department of operating theater in "Habib Bourguiba" hospital in Tunisia. We report the results obtained on three departments of different sizes. The comparative **Table 3** shows the relevant parameters of scale for the three departments.

The table shows the number of professionals for each discipline (D1: surgery, D2: anesthesia, D3: instruments) in the second column. The third column lists the possible size of team for each department which depends on the nature of the operation.

During 3 months, the team performance of the first support for building multidisciplinary teams is identified by seven tests. Respectively, two tests in orthopedics department, three tests in urology department, and two tests in the neurology department. **Table 4** shows respectively the team size for each test and size of each discipline.

Then, we apply the second support for selection team. Within our framework of aid to the choice of the best team which satisfies the preferences of decision maker and operation need. Our case base is formed by 20 operations which satisfied this type of operation.

Our objective consists on searching the best team of a new case arising to the case base. This new case is described by the same attributes that those of the other cases in base, described in **Table 5**.

The objective of similarity measures is to look for the nearest case which satisfies the most preferences of the new operation in the case base. Indeed, by applying Eq. (1), we calculate all local similarities between attributes (**Table 6**).

The relative importance weighting attributes obtained by AHP method, W_i, are listed in **Table 7**.

Department	Nb professionals	D1	D2	D3	Team size
1	22	8	8	6	5-6
2	36	10	12	14	4-5-6
3	42	12	15	16	6-7

Table 3.
The experimental departments.

Test	Team size	Size of each discipline		
		D1	D2	D3
1	5	2	2	1
2	6	3	2	3
3	4	2	1	1
4	5	2	1	2
5	6	2	2	2
6	6	3	2	1
7	7	3	3	1

Table 4.
Tests of proposed model.

Case	Criteria				Team
	T	CT	CO	R	
1	125	4	6	1	$\{C_2, C_3, A_5, I_2, I_3\}$
2	122	5	3	2	$\{C_4, C_1, A_2, I_1, I_4\}$
3	130	5	5	3	$\{C_6, C_3, A_6, I_2, I_5\}$
4	110	4	5	2	$\{C_{10}, C_2, A_5, I_{12}, I_3\}$
5	160	4	4	5	$\{C_4, C_2, A_2, I_3, I_2\}$
6	74	3	6	2	$\{C_1, C_7, A_{10}, I_6, I_{14}\}$
7	115	6	5	3	$\{C_3, C_6, A_3, I_9, I_{10}\}$
8	65	5	4	1	$\{C_5, C_8, A_1, I_8, I_7\}$
9	85	2	3	1	$\{C_2, C_1, A_3, I_{12}, I_3\}$
10	75	4	5	3	$\{C_7, C_5, A_8, I_8, I_{11}\}$
11	100	6	5	2	$\{C_4, C_3, A_2, I_{10}, I_2\}$
12	92	4	6	3	$\{C_9, C_5, A_7, I_9, I_5\}$
13	122	3	4	3	$\{C_6, C_{10}, A_3, I_5, I_6\}$
14	160	5	4	5	$\{C_6, C_2, A_{10}, I_{12}, I_5\}$
15	125	3	6	4	$\{C_3, C_{10}, A_2, I_{14}, I_9\}$
16	140	1	3	1	$\{C_2, C_4, A_6, I_3, I_6\}$
17	76	3	4	1	$\{C_5, C_9, A_1, I_9, I_{15}\}$
18	85	2	3	1	$\{C_3, C_6, A_9, I_{13}, I_2\}$
19	134	5	2	2	$\{C_8, C_1, A_3, I_{15}, I_3\}$
20	124	3	5	2	$\{C_2, C_4, A_2, I_4, I_{10}\}$
C^{New}	120	5	5	3	?

Table 5.
Case base construction for the team selection problems.

The attribute weights are then employed in Eq. (2) to measure the similarity between the cases in memory and the new case. Next, we obtain the resul in **Table 8**.

Case	T	CT	CO	R
1	0.9473	0.8	0.75	0.5
2	0.0210	1	0.5	0.75
3	0.8947	1	1	1
4	0.8947	0.8	1	0.75
5	0.5789	0.8	0.75	0.5
6	0.5157	0.6	0.75	0.75
7	0.9473	0.8	1	1
8	0.4210	1	0.75	0.5
9	0.6315	0.4	0.5	0.5
10	0.5263	0.8	1	1
11	0.7894	0.8	1	0.75
12	0.7052	0.8	0.75	1
13	0.9789	0.6	0.75	1
14	0.5789	1	0.75	0.5
15	0.9473	0.6	0.75	0.75
16	0.7894	0.2	0.5	0.5
17	0.5368	0.6	0.5	0.5
18	0.6315	0.4	0.5	0.5
19	0.8526	1	0.25	0.75
20	0.9578	0.6	1	0.75

Table 6.
Similarities local calculation.

Attributes	T	CT	Co	R	Weight (Wi)
T	0.1	0.086	0.076	0.120	0.095
CT	0.3	0.260	0.307	0.240	0.276
Co	0.2	0.130	0.153	0.159	0.160
R	0.4	0.521	0.461	0.480	0.465

Table 7.
Attributes weight.

The computational study pretends to analyze if the model improves the effectiveness of the team in operating theater and how good is its contribution. For this study, the team performances are identified by 30 tests. Respectively, 10 tests in orthopedics department, 10 tests in urology department, and 10 tests in the neurology department.

Finally, to assess the efficiency of our proposed model, we used model in the three departments of Habib Bourguiba hospital and we obtained the percentage of operation success in each department (see **Figure 6**). It analyzes the comparison of results before and after the integration of our model.

Case	Global similarities	Rank
1	0.3065	15
2	0.4065	11
3	0.9859	1
4	0.5367	7
5	0.3058	16
6	0.7178	4
7	0.8706	2
8	0.3177	13
9	0.2791	18
10	0.8595	3
11	0.5354	8
12	0.6232	5
13	0.6156	6
14	0.3129	14
15	0.4187	10
16	0.2753	20
17	0.2830	17
18	0.2791	19
19	0.3939	12
20	0.5118	9

Table 8.
Global similarities calculation.

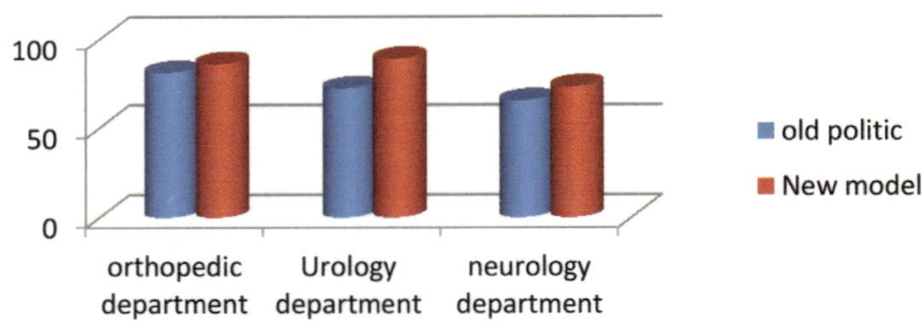

Figure 6.
Percentage of successful operations.

5. Conclusion

The proposed team building-selection model makes up for some shortages of previous models. An important contribution of this chapter is to bring in a practical case a theoretical modeling effort to describe a complex environment of health care services. The use of the first part of approach has allowed us to obtain high-quality solutions in very short commuting times, in spite of the size of the problem and the complexity of data and objectives. In the second part of approach, we present a team selection method based on a multicriteria aid model using case-based reasoning technique.

The proposed model was tested on the real datasets collected from the "Habib Bourguiba" Hospital in Tunisia. However, because of the nature of the information and the difficulty of obtaining the data, the number of available data points was limited. The developed model is highly representative of the reality because it uses the last experience case that satisfies the most the decision maker preferences.

The next step in our work will be the use of our approach in other areas. We are also planning to imbed this model in a general project management system that we are currently developing. The model can be improved by adding other attributes (experience, leadership, etc.) which can be studied in the future.

Author details

Ikram Khatrouch[1,2]*, Lyes Kermad[1], Anderrahman El Mhamedi[1] and Younes Boujelbene[3]

1 IUT Montreuil, University of Paris 8, France

2 University of Lyon, Saint Etienne, France

3 FSEGS, University of Sfax, Tunisia

*Address all correspondence to: Ikram.khatrouch@univ-st-etienne.fr

References

[1] Michel P, Quenon JL, Djihoud A, Tricaud-Vialle S, de Sarasqueta AM. French national survey of inpatient adverse events prospectively assessed with ward staff. Quality & Safety in Health Care. 2007;**16**:369-377

[2] Leape LL, Brennan TA, Laird N, Lawthers AG, Localio AR, Barnes BA. The nature of adverse events in hospitalized patients. Results of the Harvard Medical Practice Study II. The New England Journal of Medicine. 1991; **324**:377-384

[3] Wilson RM, Harrison BT, Gibberd RW, Hamilton JD. An analysis of the causes of adverse events from the Quality in Australian Health Care Study. The Medical Journal of Australia. 1999; **170**:411-415

[4] Atul A, Gawande MD, Michael J, Zinner MD, David M, Studdert L, et al. Analysis of errors reported by surgeons at three teaching hospitals. Journal of Surgery. 2003;**133**:614-621

[5] Fleming M, Smith S, Slaunwhite J. Investigating interpersonal competencies of cardiac surgery teams. Canadian Journal of Surgery. 2006; **49**(1):22-30

[6] The Joint Commission Sentinel Event Statistics: as of January 26, 2011. http://www.jointcommission.org/topics/default.aspx?k=795. 20 Mar 2011

[7] Fuchshuber PR, Greif W, Tidwell CR, Klemm MS, Frydel C, Wali A, et al. Power of the national surgical quality improvement program—Achieving a zero pneumonia rate in general surgery patients. The Permanente Journal. 2012; **16**(1):156-168

[8] Watson L, Pathiraja F, Depala A, O'Brien B, Beyzade S. Ensuring safe communication in health care: A response to Johnston et al on their paper 'Smartphones let surgeons know WhatsApp: An analysis of communication in emergency surgical teams'. American Journal of Surgery. 2016;**211**(1):302-308

[9] Taylor RS, Brown A, Ebrahim S, Jolliffe J, Noorani H, Rees K, et al. Exercise-based rehabilitation for patients with coronary heart disease: Systematic review and meta-analysis of randomized controlled trials. The American Journal of Medicine. 2004;**116**(10):682-692

[10] Taylor RS, Brown A, Ebrahim S, Jolliffe J, Noorani H, Rees K, Skidmore B, Stone JA, Thompson DR, Oldridge N. Exercise-based rehabilitation for patients with coronary heart disease: systematic review and meta-analysis of randomized controlled trials. American Journal of Medicine. 2004 May 15;**116**(10):682-692

[11] Jiang J, Zhang YM. Assessment of the effect of team trust on GDISD team efficiency based on the AHP method. Journal of Information, Computer and Application Engineering. 2015:**15**:463-467

[12] Strnad D, Guid N. A fuzzy-genetic decision support system for project team formation. Journal of Applied Soft Computing. 2010;**10**:1178-1187

[13] Ahmed F, Deb K, Jindal A. Multi-objective optimization and decision making approaches to cricket team selection. Journal of Applied Soft Computing. 2013;**13**:402-414

[14] Sahin YG. A team building model for software engineering courses term projects. Journal of Computers & Education. 2011;**56**:916-922

[15] Saaty TL. Decision Making for Leaders. Belmont, California: Lifetime Learning Publications; 1980